T0231844

# BIORESPONSIVE POLYMERS

*Design and Application in Drug Delivery*

# BIORESPONSIVE POLYMERS

*Design and Application in Drug Delivery*

*Edited by*
**Deepa H. Patel, PhD**

Apple Academic Press Inc.
4164 Lakeshore Road
Burlington ON L7L 1A4
Canada

Apple Academic Press Inc.
1265 Goldenrod Circle NE
Palm Bay, Florida 32905
USA

First issued in paperback 2021

ISBN-13: 978-1-77188-855-4 (hbk)
ISBN-13: 978-1-77463-897-2 (pbk)
ISBN-13: 978-0-42932-524-3 (eBook)

### Library and Archives Canada Cataloguing in Publication

Title: Bioresponsive polymers : design and application in drug delivery / edited by Deepa H. Patel, PhD

Names: Patel, Deepa H., editor.

Description: Includes bibliographical references and index.

Identifiers: Canadiana (print) 20200202871 | Canadiana (ebook) 20200202928 | ISBN 9781771888554 (hardcover) | ISBN 9780429325243 (ebook)

Subjects: LCSH: Polymeric drug delivery systems. | LCSH: Biomimetic polymers.

Classification: LCC RS199.5 .B56 2020 | 615/.6—dc23

### Library of Congress Cataloging-in-Publication Data

Names: Patel, Deepa H., editor.

Title: Bioresponsive polymers : design and application in drug delivery / edited by Deepa H. Patel.

Description: Burlington, ON ; Palm Bay, Florida : Apple Academic Press, [2021] | Includes bibliographical references and index. | Summary: "This volume, Bioresponsive Polymers: Design and Application in Drug Delivery, focuses on recent advancements in bioresponsive polymers and their design, characterization, and applications in varied fields, such as drug delivery and gene delivery. It looks at several carriers for drug delivery and biological molecules using different bioresponsive polymers. To address the many difficulties in existing dosage forms, this book provides information on recent developments to overcome drawbacks of conventional forms of the drug delivery. The chapters cover most areas of bioresponsive polymers, starting with a basic introduction to bioresponsive polymers, followed by chapters on design, characterization, and mechanism of bioresponsive polymers; and applications of drug and gene delivery using bioresponsive polymers via oral, topical, nasal, ocular, and parenteral methods. The book also reviews recent advancements in bioresponsive polymers and advanced applications, such as engineering particulate moieties, biomedical applications, hydrogels as emerging therapy, and electrochemical responses, bioresponsive nanoparticles, and bioresponsive hydrogels. This book will be valuable to researchers of various fields of bioresponsive polymers along with faculty and students in this area"-- Provided by publisher.

Identifiers: LCCN 2020012226 (print) | LCCN 2020012227 (ebook) | ISBN 9781771888554 (hardcover) | ISBN 9780429325243 (ebook)

Subjects: LCSH: Polymeric drug delivery systems. | Polymeric drugs. | Drugs--Design. | Biomedical materials.

Classification: LCC RS201.P65 B58 2021  (print) | LCC RS201.P65  (ebook) | DDC 615.1/9--dc23

LC record available at https://lccn.loc.gov/2020012226

LC ebook record available at https://lccn.loc.gov/2020012227

Apple Academic Press also publishes its books in a variety of electronic formats. Some content that appears in print may not be available in electronic format. For information about Apple Academic Press products, visit our website at **www.appleacademicpress. com** and the CRC Press website at **www.crcpress.com**

# About the Editor

---

**Deepa H. Patel, PhD, MPharm**

*Associate Professor, Department of Pharmaceutics, Faculty of Pharmacy, Parul Institute of Pharmacy and Research, Parul University, Vadodara–391760, Gujarat, India, Phone: 02668-260287, Fax: 02668-260201, E-mails: deepaben.patel@paruluniversity.ac.in, pateldeepa18@yahoo.com*

Deepa H. Patel, MPharm, PhD, is Associate Professor in the Department of Pharmaceutics at the Parul Institute of Pharmacy and Research in the Faculty of Pharmacy at Parul University, Vadodara, India. She has written several book chapters and papers published in professional journals. She has filed four Indian patent applications. Dr. Patel has received many awards for her research and teaching work as well as several research grants and fellowships. Her experience is in teaching and research as well as in quality assurance of pharmaceutical products. Her research focuses on developing and translating new drug delivery systems for the prevention and treatment of various diseases, such as cancer, HIV, rheumatoid arthritis, inflammation, and asthma, as well as particle engineering using various commercially available techniques, such as spray drying, supercritical fluid particle formers, and lyophilizers.

# Contents

*Contributors* ........................................................................................ *ix*

*Abbreviations* ...................................................................................... *xi*

*Preface* ............................................................................................... *xix*

1. **Introduction to Bioresponsive Polymers** ................................................ 1
   Deepa H. Patel, Drashti Pathak, and Neelang Trivedi

2. **Design of Bioresponsive Polymers** ........................................................ 41
   Anita Patel, Jayvadan K. Patel, and Deepa H. Patel

3. **Application of Bioresponsive Polymers in Drug Delivery** ..................... 73
   Manisha Lalan, Deepti Jani, Pratiksha Trivedi, and Deepa H. Patel

4. **Application of Bioresponsive Polymers in Gene Delivery** ................... 121
   Tamgue Serges William, Drashti Pathak, and Deepa H. Patel

5. **Recent Developments in Bioresponsive Drug Delivery Systems** ......... 151
   Drashti Pathak and Deepa H. Patel

6. **Bioresponsive Nanoparticles** ............................................................... 173
   Drashti Pathak and Deepa H. Patel

7. **Bioresponsive Hydrogels for Controlled Drug Delivery** ..................... 197
   Tamgue Serges William, Dipali Talele, and Deepa H. Patel

*Index* ............................................................................................... *233*

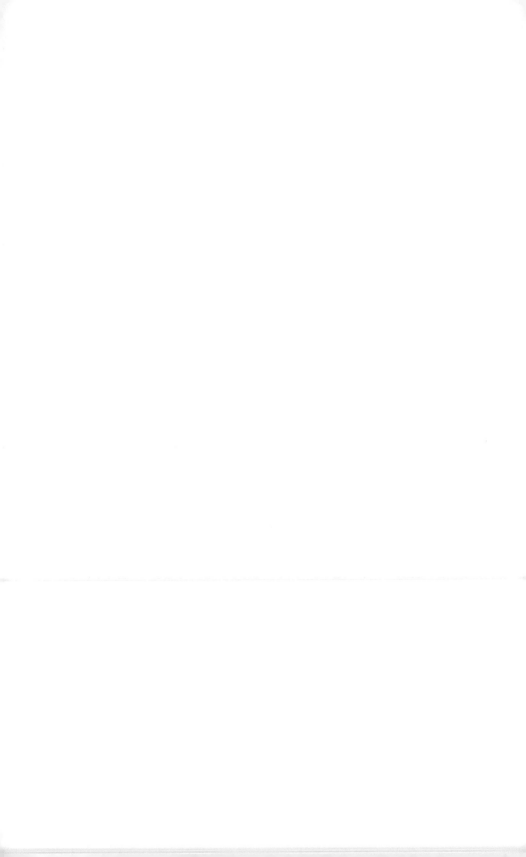

# Contributors

**Deepti Jani**
Assistant Professor, Department of Pharmaceutics, Babaria Institute of Pharmacy, NH#8,
PO. Varnama, Distt. Vadodara–391247, Gujarat, India

**Manisha Lalan**
Associate Professor, Department of Pharmaceutics, Babaria Institute of Pharmacy, NH#8,
PO. Varnama, Distt. Vadodara–391247, Gujarat, India, E-mail: manisha_lalan79@yahoo.co.in

**Anita Patel**
Assistant Professor and HOD, Nootan Pharmacy College, Faculty of Pharmacy,
Sankalchand Patel University, Visnagar–384315, Gujarat, India

**Deepa H. Patel**
Associate Professor, Department of Pharmaceutics, Parul Institute of Pharmacy and Research,
Faculty of Pharmacy, Parul University, P.O. Limda, Ta. Waghodia, Dist. Vadodara–391760,
Gujarat, India, Phone: 02668-260287, Fax: 02668-260201,
E-mails: deepaben.patel@paruluniversity.ac.in, Pateldeepa18@yahoo.com

**Jayvadan K. Patel**
Principal and Dean, Nootan Pharmacy College, Faculty of Pharmacy, Sankalchand Patel University,
Visnagar–384315, Gujarat, India, E-mail: jayvadan04@gmail.com

**Drashti Pathak**
Assistant Professor, Department of Pharmaceutics, Parul Institute of Pharmacy and Research,
Faculty of Pharmacy, Parul University, P.O. Limda, Ta. Waghodia, Dist. Vadodara–391760,
Gujarat, India

**Dipali Talele**
Assistant Professor, Department of Pharmaceutics, Parul Institute of Pharmacy and Research,
Faculty of Pharmacy, Parul University, P.O. Limda, Ta. Waghodia, Dist. Vadodara–391760,
Gujarat, India

**Neelang Trivedi**
Research Scholar, Department of Pharmaceutics, Parul Institute of Pharmacy and Research,
Faculty of Pharmacy, Parul University, P.O. Limda, Ta. Waghodia, Dist. Vadodara–391760,
Gujarat, India

**Pratiksha Trivedi**
Assistant Professor, Department of Pharmaceutics, Babaria Institute of Pharmacy, NH#8,
PO. Varnama, Distt. Vadodara–391247, Gujarat, India

**Tamgue Serges William**
Research Student, Department of Pharmaceutics, Parul Institute of Pharmacy and Research,
Faculty of Pharmacy, Parul University, P.O. Limda, Ta. Waghodia, Dist. Vadodara–391760,
Gujarat, India

# Abbreviations

| | |
|---|---|
| AAV | adeno-associated virus |
| ABP | arginine-grafted poly(cystaminebisacrylamide diaminohexane) |
| ACPPs | activatable CPPs |
| ADP | adenosine diphosphate |
| AEA | polyvinylacetal diethylaminoacetate |
| AMD | age-related macular degeneration |
| APy | n-acryloyl pyrrolidine |
| ASGP | asialoglycoprotein |
| ATP | adenosine triphosphate |
| ATR | attenuated total reflection |
| ATRP | atom transfer radical polymerization |
| AV | adenoviruses |
| AZO | azobenzene |
| BA | boronic acid |
| BACy | bisacryloylcystamine |
| b-CD | beta cyclodextrin |
| BCNU | 1,3-bis (2-chloroethyl)-1-nitrosourea |
| BDDGE | 1,4-butanediol diglycidylether |
| BDNF | brain-derived neurotrophic factor |
| BGLs | blood glucose levels |
| B-PDEAEA | poly [(2-acryloyl) ethyl (p-boronic acid benzyl) diethylammonium bromide] |
| BSA | bovine serum albumin |
| CARs | chimeric antigen receptors |
| CAT | catalase |
| CBA | cystaminebisacrylamide |
| CCP | charge conversion property |
| CC-RRNs | core-crosslinked redox responsive nanoparticles |
| CMC | carboxymethyl cellulose |
| CM-Dex | carboxymethyl dextran |
| CMT | critical micellization temperature |
| CMV | cytomegalovirus |

| | |
|---|---|
| Con A | concanavalin A |
| CRPs | controlled radical polymerizations |
| Cryo-TEM | cryogenic transmission electron microscopy |
| CS | chitosan |
| CS-PLGA | chitosan modified PLGA |
| CTA | chain transfer agent |
| CTPP | (3-carboxypropyl) triphenyl phosphonium bromide |
| Cys | cysteine |
| DCs | dendritic cells |
| DHP | diamminedichlorodihydroxy-platinum |
| DMA | dynamic mechanical analysis |
| DMAEMA | dimethylaminoethyl methacrylate |
| DMMAn | dimethylmaleic anhydride-modified melittin |
| DMPC | 1,2-dimyristoyl-sn-glycero-3-phosphorylcholine |
| DNA | deoxyribonucleic acid |
| DOX | doxorubicin |
| DPPC | dipalmitoylphosphatidylcholine |
| DSC | differential scanning calorimetry |
| DSP | diamminedichlorodisuccinato-platinum |
| DSPC | 1,2-distearoyl-sn-glycero-3-phosphorylcholine |
| DTT | dithiothreitol |
| DTX | docetaxel |
| EBV | Epstein Barr virus |
| ECH-HPP | epichlorohydrin |
| ECM | extracellular matrix |
| EGDMA | ethylene glycol dimethacrylate |
| EGFP | enhanced green fluorescent protein |
| EGF-PEG | epidermal growth factor-poly ethylene glycol |
| EGFR | epithelial growth factor receptor |
| EMHPs | elastin-mimetic hybrid polymers |
| EMMA | endosomolysis by masking of a membrane-active agent |
| EMR | electromagnetic radiations |
| EOEOVE | 2-(2-ethoxy) ethoxyethyl vinyl ether |
| EOn-POm-EOn | assembled units of ethylene oxide and propylene oxide |
| EPR | enhanced permeability and retention |

| | |
|---|---|
| ERSPC | European Randomized Study of Screening for Prostate Cancer |
| ESF | European Science Foundation |
| FDA | Food and Drug Administration |
| FEM | finite element modeling |
| FITC-dextran | fluorescein isothiocyanate dextran |
| FTIR | Fourier transform infrared |
| GAG | glycosaminoglycan |
| GAPDH | glyceraldehyde 3-phosphate dehydrogenase |
| GLP-1 | glucagon-like peptide-1 |
| GOx | glucose oxidase |
| GPC | gel permeation chromatography |
| gPGA-NPs | poly-g-glutamic acid nanoparticles |
| GRIDS | glucose-responsive insulin delivery system |
| GSH | glutathione |
| GyrB | gyrase subunit B |
| $H_2O_2$ | hydrogen peroxide |
| HA | hyaluronic acid |
| HAase | hyaluronidase |
| HaEMA | hydroxyethyl methacrylate |
| HBV | hepatitis B virus |
| HDL | high-density lipoproteins |
| HECHA | hydroxyethyl cellulose/hyaluronic acid |
| HEMA | hydroxyethyl methacrylate |
| HIV | human immunodeficiency virus |
| HLB | hydrophilic-lipophilic-balance |
| HMW | high molecular weight |
| HNE | human neutrophil elastase |
| HOCl | hypochlorous acid |
| Hp-b-CD | hydroxypropyl-b-cyclodextrin |
| HPMA | N-(2-hydroxypropyl) methacrylamide |
| HPMC | hydroxyl propyl methyl cellulose |
| HR-NPs | hypoxia-responsive nanoparticles |
| HSA | human serum albumin |
| HSV | herpes simplex virus |
| HUVECs | human umbilical vein endothelial cells |
| Hyal-1 | hyaluronidase-1 |
| IPNs | interpenetrating polymer networks |

| | |
|---|---|
| IR | infrared |
| ISPs | immunestimulatory peptides |
| LCST | lower critical solution temperature |
| LDL | low-density lipoproteins |
| LFUS | low frequency ultrasound |
| L-PEIs | linear PEI |
| LRP | living radical polymerization |
| MAA | maleic acid amide |
| MALLS | multi angle laser light scattering |
| mcDNA | DNA minicircles |
| MDGI | mammary-derived growth inhibitor |
| MEMS | micro electro mechanical system |
| MES | morpholinoethane sulfonic acid |
| miRNA | microRNA |
| MMP | matrix metalloproteinase |
| MMP2 | matrix metalloproteinase-2 |
| MMS | magnetic mesoporous silica |
| MN-CS-PLGA | mannan and chitosan co-modified PLGA |
| MN-PLGA | mannan modified PLGA |
| MoMLV | moloney murine leukemia virus |
| MP | methyl prednisolone |
| MPC | methacryloyloxyethylphosphorylcholine |
| MPEG-PAE | methyl ether poly(ethylene glycol) poly (beta-amino ester) block copolymer |
| mPEG-PS | methoxy polyethylene glycol-b-poly (diethyl sulfide) |
| mPEG-TK-PCL | methoxy (polyethylene glycol) thioketal-poly($\varepsilon$-caprolactone) |
| MPS | mononuclear phagocytic system |
| MRI | magnetic resonance imaging |
| MRSA | methicillin-resistant staphylococcus aureus |
| MSPs | mesoporous silica nanoparticles |
| MTX | methotrexate |
| NCA | N-carboxy anhydride |
| NHS | N-hydroxysuccinimide |
| NI | nitroimidazole |
| NIPAMM | N-isopropylacrylamide |
| NIR | near-infrared |

| | |
|---|---|
| NMP | N-methyl-2-pyrrolidone |
| NMR | nuclear magnetic resonance |
| NMRP | nitroxide mediated radical polymerization |
| NO | nitric oxide |
| NP | nanoparticle |
| NVP | N-vinyl-2-pyrrolidone |
| O | oxygen |
| ODNs | oligodeoxinucleotides |
| OH | hydroxyl radical |
| ONOO- | peroxynitrite |
| P(Asp-Az)X-SS-PEIs | polyaspartamide-based disulfide-containing brushed polyethylenimine |
| P(Asp-DIP) | poly(N-(N',N'-diisopropylaminoethyl) aspartamide) |
| P(IPAAm-co-DMAAm) | poly(N-isopropylacrylamide-co-N, N-dimethylacrylamide) |
| P(Lys-Ca) | poly(lysine-cholic acid) |
| P(MAA-co-NVP) | poly(methacrylic acid-co-n-vinyl pyrrolidone) |
| P(MAA-g-EG) | poly(methacrylic acid-g-poly(ethylene glycol)) |
| P(VCL-s-s-MAA) | poly(N-vinylcaprolactam-co-methacrylic acid) |
| PA6ACA | acryloyl 6-aminocaproic acid |
| PAA | poly(acrylic acid) |
| PAE | poly($\beta$-amino ester) |
| PAMAM | poly(amidoamine) |
| PAMPs | pathogen-associated molecular patterns |
| PAMPS | poly(2-acrylamido-2-methylpropane sulfonic acid) |
| PAP | PAMAM-azo-peg |
| PAsp | polyaspartamide |
| PBA | phenyl boronic acid |
| PBAE | poly($\beta$-amino ester) |
| PBS | phosphate-buffered saline |
| PCCA | polymerized crystalline colloidal array |
| PCLDMA | poly($\varepsilon$-caprolactone) dimethacrylate |
| PD-1 | programmed death-1 |
| PDAAEMA | poly(N, N-diakylaminoethylmethacrylates) |
| pDbB | poly(dimethylaminoethylmethacry late-block-butylmethacrylate) |
| PDEA | poly[(2-diethylamino) ethyl methacrylate] |

| | |
|---|---|
| PDEAEMA | poly(*N*, *N*-diethylaminoethylmethacrylate) |
| PDI | polydispersity index |
| PDMA | poly {(2-dimethylamino)ethyl methacrylate |
| PDMA-b-PDPA | poly(2(dimethylamino)ethyl methacrylate) blockpoly(2(diisopropylamino)ethyl methacrylate) |
| PDMAEMA | poly(dimethyl amino ethyl methacrylate) |
| PDMS-b-PDMAEMA | poly(dimethysiloxane)-b-poly(2-(dimethylamino)ethyl methacrylate) |
| pDNA | plasmid DNA |
| PDPA | poly[(2-diisopropylamino) ethyl methacrylate] |
| PDT | photodynamic therapy |
| PEG | polyethylene glycol |
| PEG2k-PCL5k | poly(ethylene glycol)-block-polycaprolactone |
| PEG-a-PDMAEMA | PEG and poly(2-(dimethylamino)ethyl methacrylate) |
| PEG-b-P(AlA-co-MAA) | poly(ethylene glycol)-b-poly alkyl acrylate-co-methacrylic acid) |
| PEG-b-PAsp(DET) | polyethylene glycol-b-polyaspartamide diamioethane |
| PEG*b*PHF | poly(ethylene glycol)-b- poly (L-histidine-co-L-24 phenylalanine) |
| PEG-b-PLLA | poly(ethylene glycol)-block-poly(d,l-lactic acid) |
| PEG-SHGel | PEG modified thiolated gelatin |
| PEI | polyethylenimine |
| PEI-NI | polyethyleneimine-nitroimidazole micelles |
| PEI-SS-CLs- | disulfide-containing cross-linked polyethylenimines |
| PELT | polymer enzyme liposome therapy |
| PEO | PAAc-polyethylene oxide |
| PEO | polyethylene oxide |
| PEO-*b*-pNIPAm | poly(ethylene oxide)-*block*-poly(*N*-isopropylacrylamide) |
| PEO-PPO-PEO | poly(ethylene glycol)-*block*-poly(propylene glycol)-*block*-poly(ethylene glycol) |
| PEOz-PLA-g-PEI-SS | poly(2-ethyl-2-oxazoline)- poly(L-lactide) grafted with bioreducible polyethylenimine |
| pHEMA | polyhydroxyethylmethacrylate |
| PHPMA | N-(2-hydroxypropyl) methacrylamide |

| | |
|---|---|
| PKCα | protein kinase Cα |
| PL | polylysine |
| PLA2 | phospholipase A2 |
| PLCO | prostate, lung, colorectal, and ovarian cancer screening trial |
| PLG | poly-L-glutamate |
| PLGA | poly(lactic-co-glycolic acid) |
| PLK-1 | polo-like kinase 1 |
| PLL | poly(L-lysine) |
| PLL-g-Chi | poly(L-lysine)-graft-chitosan |
| PMA | poly(MPC-co-methacrylic acid) |
| PMAA | poly methacrylic acid |
| PMAA-PMA | poly(methacrylic acid) polymethacrylate copolymer |
| PMEMA | poly[(2-N-morpholino) ethyl methacrylate] |
| PMN | polymorphonuclear |
| PMPC | PEG-b- poly(MPC)n |
| PNIPAAM | poly(N-isopropylacrylamide) |
| PNVCL | poly(n-vinyl caprolactam) |
| poly(AA)-g-PEG | **poly**(acrylic acid)**-graft-poly**(ethylene glycol) |
| poly(IA-co-NVP) | poly(itaconic acid-co-N-vinyl-2-pyrrolidone) |
| poly(NIPAAM) | poly(N-isopropylacrylamide) |
| PPADT | poly-(1, 4-phenyleneacetone dimethylene thioketal) |
| PPS | poly(propylene sulfide) |
| PRINT | particle replication in non-wetting templates |
| PSA | prostate-specific antigen |
| pSMA | poly(styrene-alt-maleic anhydride) |
| PSSA | poly(4-styrenesulfonic acid) |
| PTX | paclitaxel |
| PVA | polyvinyl alcohol |
| PVBPA | poly(4-vinyl-benzyl phosphoric acid) |
| PVP | poly vinyl pyrrolidone |
| PVPA | poly(vinylphosphoric acid) |
| PVSA | poly(vinylsulfonic acid) |
| QELS | quasi-elastic light scattering |
| RAFT | reversible addition fragmentation transfer |
| RBCs | red blood cells |

| | |
|---|---|
| RES | reticuloendothelial system |
| rHB | reducible hyperbranched |
| RHB | reducible hyperbranched poly-(amidoamine) |
| RNAi | RNA interference |
| RO | alkoxy radicals |
| ROMP | ring-opening metathesis polymerization |
| ROS | reactive oxygen species |
| SANS | small-angle neutron scattering |
| SAS | small-angle scattering |
| SEM | scanning electron microscope |
| Si | silicon |
| siRNA | small interfering RNA |
| SOD | superoxide dismutase |
| SPAnH | poly [aniline-co-n-(1-butyric acid) aniline] |
| sPLA$_2$ | secretary phospholipase a$_2$ |
| SRNs | stimuli responsive nanocarriers |
| -SS- | disulfide bond |
| SWNTs | single-walled carbon nanotubes |
| TKNs | thioketal nanoparticles |
| TMBQ | trimethyl-locked benzoquinone |
| TMC-g-PNIPAAm | trimethyl chitosan-grafted by poly(N-isopropylacrylamide) |
| TME | tumor microenvironment |
| TNF-$\alpha$ | tumor necrosis factor-alpha |
| TRAIL | tumor necrosis factor-related apoptosis inducing ligand |
| TRITC-Con A | tetramethylrhodamineisothiocyanateconcavalin A |
| Trx | thioredotoxin |
| UCST | upper critical solution temperature |
| UV | ultraviolet |
| VAc | vinyl acetate |
| VEGF | vascular endothelial growth factor |
| W/O | water-in-oil |
| $\beta$-gal | $\beta$-galactosidase |

# Preface

The purpose of this book is to focus on bioresponsive polymers and their design, characterization, and applications in different fields such as drug delivery and gene delivery. Moreover, the present book also focuses on recent advancements and several carriers to deliver drug and biological molecules using different bioresponsive polymers.

The intention to write this book is to target various researchers such as academicians and industry persons who are associated with bioresponsive polymer-based research. In the current era, we are facing many difficulties in existing dosage forms. This book presents recent developments to overcome drawbacks of the conventional form of drug delivery. Many research scholars, as well as professors, have started their research on bioresponsive polymers, so this book will be an important resource to get information related to the bioresponsive polymer as an emerging trend.

The book mainly has seven chapters that cover almost all areas of bioresponsive polymers. Chapter 1 provides a basic introduction regarding bioresponsive polymers followed by six more chapters that include design, characterization, and mechanism of bioresponsive polymers as well as applications of drug and gene delivery using bioresponsive polymers via oral, topical, nasal, ocular, and parenteral means. Furthermore, the chapters also address recent advancements in bioresponsive polymers and their advanced applications, such as engineering particulate moieties, biomedical applications, hydrogels as emerging therapy, and electrochemical response. The present book also covers the details of bioresponsive nanoparticles and hydrogels.

We sincerely hope the present book will receive an overwhelming response from researchers from various fields of bioresponsive polymers.

We thank and acknowledge the publishers, all our book contributors, and our families for their direct and indirect support to complete this important task. We would also like to thank all staff of Apple Academic Press for shaping this book in the emerging field of bioresponsive polymers.

—**Deepa H. Patel, PhD**

# CHAPTER 1

# Introduction to Bioresponsive Polymers

DEEPA H. PATEL, DRASHTI PATHAK, and NEELANG TRIVEDI

*Department of Pharmaceutics, Parul Institute of Pharmacy and Research, Faculty of Pharmacy, Parul University, P.O. Limda, Ta. Waghodia, Vadodara–391760, Gujarat, India, Phone: 02668-260287, Fax: 02668-260201, E-mails: deepaben.patel@paruluniversity.ac.in, Pateldeepa18@yahoo.com (D. H. Patel)*

## 1.1 HISTORY

Smart' bioresponsive materials that are sensitive to biological signals or to pathological abnormalities, and interact with or are actuated by them, are appealing therapeutic platforms for the development of next-generation precision medications. Armed with a better understanding of various biologically responsive mechanisms, researchers have made innovations in the areas of materials chemistry, biomolecular engineering, pharmaceutical science, and micro- and nanofabrication to develop bioresponsive materials for a range of applications, including controlled drug delivery, diagnostics, tissue engineering, and biomedical devices.

In the last decade, the area of drug delivery has concentrated on two criteria: (1) controlling the release of a therapeutic agent; and (2) targeting of the therapeutic agent to a specific target site. These research accomplishments have focused mostly on the development of novel biodegradable polymers and target moiety-labeled drug delivery carriers. The latest interests in biomaterials that act in response to their environment have opened new systems to activate the release of drugs and localize the drug within a target site. These modern biomaterials, generally termed "smart" or "intelligent," are proficient to deliver a therapeutic agent based on either environmental signals or a remote stimulus. Bioresponsive materials could potentially elicit a therapeutically effective dose without any side effects. Responding of polymers to different stimuli, such as pH, light, temperature, ultrasound, magnetism, or biomolecules have been explored as potential drug delivery vehicles. This chapter describes the most recent advances in "smart" drug delivery systems that respond to one or multiple stimuli [1].

Polymers that respond to a stimulus are often called "smart" or "intelligent" owing to their inherent ability to modify their physical or chemical properties. For the majority of the polymers that fall into this class, the response to a alter in the surrounding environment is not only rapid, on the command of minutes to hours, but also reversible, imitating the dynamics detected in natural polymers, such as proteins, polysaccharides, and nucleic acids in living organic systems. The response to a stimulus is noticeable in many types: individual chain size, shape, surface morphology, secondary structure, solubility, and degree of intermolecular organization. These unique abilities have been useful to a various range of applications, such as drug delivery, diagnostics, biological coating technologies, biosensing, and microfluidics [1].

Traditional drug delivery system entraps the molecules physically within a polymer matrices; drug is released in a sustained manner by diffusion or by the degradation of the polymeric network. These systems typically consequence at an early peak in plasma drug concentration followed by a steady-state, linear drug release. This is far away from ultimate because the drug concentration at a specific location is not precisely controlled which causes under medication or over medication. Below the effective therapeutic dose, the drug is ineffective whereas high concentrations of the drug may cause the toxic effect or lead to undesirable adverse side effects. Polymers have been used to modify the drug release, which sustains the drug concentration within the desired therapeutic range. On the other hand, such controlled release systems are insensible to metabolic changes in the body and are incapable of neither alter drug release nor target the drug to the diseased site. This requires of control has stimulated the utilization of bioresponsive polymers as drug carriers. As before time as the 1950s, stimuli-responsive hydrogels have been, develop for drug release. Since then, polymers that act in response to different environment stimuli have been developed. These stimuli include pH, ionic strength and the incidence of metabolic chemicals such as enzymes or antigens. Such stimuli may facilitate a drug carrier to differentiate between diseased and healthy tissue. In recent times, the drug carriers that respond to magnetic fields, light, radiation, and ultrasound have also been developed. These external stimuli permit for superior control in excess of when and where the drug is released. By alteration of the formulation or chemical moieties of the polymer, the sensitivity to respond against the stimuli can be precisely controlled.

Over the last decade, bioresponsive polymeric materials have been used in biochemical sciences in numerous ways. Since the term, "smart polymeric materials" includes a wide range of diverse compounds with unique potential for biological applications, and since attention in producing and manipulating these compounds is growing up, we felt that it would be helpful to accumulate an outlook of the field to assist in catalyzing cross-fertilization of initiatives [2].

In the early 1970's natural and synthetic polymers began to be considered for controlled release formulations development. The benefits of using biodegradable polymers in scientific applications are apparent. Different types of natural and synthetic polymers are used for the development of bioresponsive drug formulation. Several researchers aim at delivering drugs to definite disease circumstances that focus on targeting carries by fastening molecules to their surfaces that exactly bind molecules up-regulated in a particular disease. Nevertheless, this advance is costly, and whether this approach actually improves delivery to anticipated locations is passionately argued. We avoid this subject by concentrating instead on the site of release, generating polymers that degrade in response to biochemical attributes of the disease state. The present book endeavors to bring together the thrilling design of these materials and the ever-growing range of their uses by concentrating on bioresponsive polymer as a key example of this technology. Bioresponsive polymers act in response to small alterations in their surroundings with remarkable changes in their physical properties [3]. Various smart polymeric drug delivery systems are shown in Table 1.1.

With the coming out of more fresh and efficient drug therapies, raised importance is being situated upon the systems by which these drugs are being directed to the body. Traditional drug delivery systems result in a peak in plasma drug concentrations followed by a plateau and finally a decline. As a result, these systems of drug delivery may achieve toxic plasma drug concentrations or ineffective plasma levels. Available marketed controlled release formulations offer various improvements, e.g., they sustain the drug within the desired therapeutic range with just a single dose therapy, they permit site-specific drug delivery which eventually may diminish the

**TABLE 1.1**  Various Smart Polymeric Drug Delivery Systems [4]

| Stimulus | Advantage | Limitation | Responsive Material |
|---|---|---|---|
| Temperature | Ease of incorporation of active moieties Simple manufacturing and formulation | – Injectability issues under application conditions.<br>– Low mechanical strength, biocompatibility issues, and instability of thermolabile drugs | – Poloxamers<br>– Poly(N-alkylacrylamide)<br>– Poly(N-vinylcaprolactam)<br><br>– Cellulose<br>– Xyloglucan<br>– Chitosan |
| pH | Suitable for thermolabile drugs | – Low mechanical strength<br>– Lack of toxicity data | – Poly(methacrylic acid)<br>– Poly(vinyl pyridine)<br>– Poly(vinyl imidazole) |
| Light | Ease of controlling the trigger mechanism Accurate control over the stimulus | – Low mechanical strength of gel, chance of leaching out of noncovalently attached Chromophores Inconsistent responses to light | – Modified poly(acrylamide)s |
| Electric field | Pulsative release with changes in electric current | – Surgical implantation required<br>– Need of an additional equipment for external application of stimulus<br>– Difficulty in optimizing the magnitude of electric current | – Sulfonated polystyrene<br>– Poly(thiophene)<br>– Poly(ethyl oxazoline) |
| Ultrasound | Controllable protein release | – Specialized equipment for controlling the release<br>– Surgical implantation required for nonbiodegradable delivery system | – Ethylene-vinyl acetate |

need for follow-up care, they protect drugs that are rapidly destroyed by the body and in the end improve the patient compliance. Even though these methods are therapeutically beneficial over the traditional systems, they hang about insensitive to the changing metabolic states of the body. Mechanisms competent in responding to these physiological alterations must be provided in order to coordinate drug release profiles with altering the physiological conditions. Preferably, a drug delivery system should respond to physiological needs, sense the alteration, and modify the drug-release profiles accordingly. This includes the two perceptions of (1) temporal modulation, and (2) drug targeting a specific site. The symptoms of most disease circumstances follow a rhythmic pattern, for instance, angina pectoris, diabetes mellitus, etc., and have need of drug delivery as per the rhythms. More importantly, if the drug possesses any adverse side effects, drug release when not needed poses an additional burden on the body's metabolic system. A more suitable and effectual approach of controlling some of these situations is by chronotherapy. This advance permits for pulsed or self-regulated drug delivery which is controlled to the staging of biological rhythms, since the onset of definite diseases exhibits strong circadian temporal dependence [3, 4].

## 1.2 BIORESPONSIVE POLYMERS

Biologically responsive polymer systems are increasingly important in various biomedical applications. The major advantage of bioresponsive polymers is that they can respond to the stimuli that are inherently present in the natural system. Bioresponsive polymeric systems mainly arise from common functional groups that are known to interact with biologically relevant species, and in other instances, the synthetic polymer is conjugated to a biological component. Bioresponsive polymers are classified into antigen responsive polymers, glucose-sensitive polymers, and enzyme responsive polymers (Table 1.2).

### 1.2.1 *pH SENSITIVE POLYMERS*

pH-responsive materials are often capable of physical or chemical changes, such as swelling, shrinking, dissociation, degradation, or membrane fusion and disruption. The pH-sensitivity can be attributed to either the protonation

**TABLE 1.2**   Summary of Typical Physiological Stimuli [5]

| Body Part or Biological Stimulus | Details |
|---|---|
| **pH** | |
| Plasma | Normal pH range: 7.38–7.42 |
| Gastrointestinal tract | Saliva: 6.0–7.0; gastric fluid: 1.0–3.5; bile: 7.8; pancreatic fluid: 8.0–8.3; small-intestinal fluid: 7.5–8.0; large-intestinal fluid: 5.5–7.0 |
| Urinary tract | Urine of pH-balanced body: 6.5–8.0 |
| Vagina | Normal pH range: 3.8–4.5 |
| Eye | Ocular surface: ~7.1; healthy tear: 7.3–7.7 |
| Pathological microenvironment | Inflammation-associated acidic pH: 7.2–6.5 for extracellular pH in tumor; down to pH 5.4 in inflamed tissue; down to 4.7 in fracture-related hematomas; down to pH 5.7 in cardiac ischemia |
| Intracellular compartments | Early endosome: 6.0–6.5; late endosome: 5.0–6.0; lysosome: 4.5–5.0; Golgi complex: 6.0–6.7 (REFS 106, 238, 239) |
| **Redox** | |
| Reducing species | Glutathione (GSH): intracellular, 10 mM; extracellular fluids, 2–10 μM |
| Oxidative species | Elevated reactive oxygen species levels are associated with inflammation and tissue injury. |
| **Enzyme** | |
| MMPs | Overexpression of MMPs is associated with various cancers and colorectal disease. For example, the plasma MMP-9 level in a healthy human body is about half of that found in patients with non-small cell lung cancer. Moreover, evidence suggests that MMPs are important regulators for inflammatory and wound-healing processes. |
| HAase | Breast cancer: elevated HAase levels in metastases compared with the primary tumor; prostate cancer: 3–10-fold elevation in HAase expression of cancerous tissue compared with that of normal adult prostate; bladder cancer: 5–8-fold higher urinary HAase levels in patients with grade 2 and 3 bladder tumor than those of normal individuals. Note: the tissue HAase levels usually correlate with tumor grade. |
| Phospholipases | Typical plasma levels of type II secretary phospholipase A2 (sPLA2) in healthy individuals are 5.8–12.6 ng ml; elevated sPLA2 levels are associated with potential artery diseases. |
| PSA | The lower threshold of PSA was established to be >4 ng ml$^{-1}$ ERSPC trial; elevated PSA levels are often associated with prostate cancer in a PLCO trial and >3 ng ml$^{-1}$ based on the ERSPC trial; elevated PSA levels are often associated with prostate cancer. |

**TABLE 1.2** *(Continued)*

| Body Part or Biological Stimulus | Details |
|---|---|
| **Glucose** | |
| Diabetic | Blood glucose level: |
| | >180 mg dl⁻¹ (hyperglycemia) |
| | <70 mg dl⁻¹ (hypoglycemia) |
| Non-diabetic | Blood glucose level: |
| | 70–100 mg dl⁻¹ (normal fasting blood glucose level); |
| | <140 mg dl⁻¹ (normal blood sugar level, 2 hours post-eating) |
| **Physical stimulus** | |
| Temperature | Normal body temperature: 36.5–37.5°C |
| Pressure and shear force | Normal range of mean arterial pressure: 70–105 mmHg. The average shear stress in healthy coronary arteries is found to be around 1.5 Pa; in constricted vessels, it increases to above 7 Pa (REF. 250) |
| **Others** | |
| ATP | Intracellular environment: 1–10 mM; extracellular environment: <5 µM (REF. 248) |
| Hypoxia | Hypoxemia (abnormally low blood oxygen level): <60 mmHg Hypoxia (low oxygen levels in tissues): critical oxygen partial pressure is 8–10 mmHg on a global tissue level; for example, regions with low oxygen partial pressures (down to zero) often exist in solid tumors |

*Abbreviations:* ATP–adenosine triphosphate; ERSPC–European randomized study of screening for prostate cancer; HAase–hyaluronidase; MMP–matrix metalloproteinase; PLCO–prostate, lung, colorectal, and ovarian cancer screening trial; PSA–prostate-specific antigen [5].

of ionizable groups or the degradation of acid-cleavable bonds. In intracellular delivery, the acidification of endosomes and their consequent union with lysosomes produce an ideal pH gradient for intracellular drug release. The pH gradients along the gastrointestinal tract facilitate the organ-specific release of orally administrated drugs at the organ level. For disease-specific controlled drug delivery, the local acidification commonly found at cancerous or inflammatory sites has also been frequently used. Classic examples are polymers polymerized from acrylic acid, methacrylic acid, maleic anhydride and N, N-dimethylamino ethyl methacrylate. The US Food and Drug Administration (FDA) approved cationic polymer, aminoalkyl methacrylate copolymer (Eudragit E), with increased solubility in acidic environments has been applied for taste masking through the suppression of burst drug release in the oral cavity. The design and selection of pH-responsive materials also depend strongly on the nature of the payload molecules. For instance, the proton pump inhibitors and some proteins are playing a crucial role in prevention from gastric degradation specifically for acid-degradable drugs. Anionic polymers comprising of carboxyl groups have a higher solubility at basic pH and thus could be used for shielding acid-sensitive drugs for intestine targeting. For illustration, microspheres consisted of poly(itaconic acid-co-N-vinyl-2-pyrrolidone) (poly(IA-co-NVP)) were assessed for the controlled release of the model protein therapeutics triggered by basic pH [5]. Polycations are particularly interesting for non-viral gene delivery due to their easy complexation with negatively charged nucleotides by electrostatic interaction. Polyethylenimine was approved by the FDA as a gold standard for the evaluation of new polymers intended for delivery of nucleic acids, even though its efficiency and safety have been exceeded by other alternatives. For instance, a high-throughput screened library of altered poly(ß-amino ester) chemistries is used to direct the building of polycations. Acetal-based acid-labile cross linkers or pendent chains of polymers were entrapped through cross-linked microgels for the pH-activated release of therapeutic protein. The particle replication in non-wetting templates (PRINT) method has been used to integrate pH-sensitive silyl ether cross-linkers into microparticles to create acid-degradable smart devices. In recent research, a doxorubicin (DOX) was attached to the side chains of poly(lactic-co-glycolic acid) (PLGA) by means of an acid-labile hydrazine linker, to avoid excretion by drug efflux pumps. Another family of pH-responsive polyion complex micelles formed through the self-assembly of oppositely charged amphiphilic block copolymers. Moreover, pH-responsive peptide amphiphiles have been used for pH-sensitive reversible self-assembly into nanofibres. Decreased pH within the endosome activated the release of entrapped DNase in a DNA-assembled nanoclew, which attacked the DNA nanoclew for the ultimate intracellular Dox delivery. In the present case, the endosome acidification provided as an indirect activate to induce drug release, which can be considered as a gradually activated system. Even though organic materials are, still mostly used, inorganic pH-responsive materials have lately appeared as substitutes for drug delivery applications, among which acid-degradable materials such as calcium phosphate and liquid metal might be appealing because of their biodegradability and the non-toxic or low-toxicity metabolism products. In recent research, dendrimer-attached platinum prodrugs were released at a specific site within tumor tissue as an outcome of amide bond cleavage within the mildly acidic environment. Recently, researchers took the benefit of stomach-specific pH-sensitive supramolecular gels that were stable in acidic environments but soluble at neutral pH values. The enteric elastomer was assembled using poly(acryloyl 6-aminocaproic acid) (PA6ACA) and poly(methacrylic acid-co-ethyl acrylate). The terminal carboxyl groups facilitated the development of intermolecular hydrogen bonds in acidic environments, resulting in an elastic water-containing supramolecular network. By contrast, in neutral environments, the supra-molecular gels undergo rapid dissociation due to the deprotonation of carboxyl groups. The system created from this smart gel demonstrated prolonged gastric retention in a pig model. Moreover, the constructions of pH-responsive hydrogels have been exploited for regenerative medicine as a smart drug-delivery system. For instance, dimethylamino ethyl methacrylate (DMAEMA)-based scaffolds able to altering oxygen and nutrient transport by expanding in an acidic environment were used to generate a pro-healing effect for tissue regeneration [5].

The triblock copolymers PEG-P(Asp-DIP)-P(Lys-Ca) (PEALCa) of polyethylene glycol (PEG), poly(N-(N,'N'-diisopropylaminoethyl) aspartamide) (P(Asp-DIP)), and poly(lysine-cholic acid) (P(Lys-Ca)) were build up as a pH-responsive drug delivery system, which can self-assemble into stable vesicles with a size around 50–60 nm, avoid uptake by the reticuloendothelial system (RES), and cover the drug in the core at neutral aqueous atmosphere such as physiological surroundings. However, the PEALCa micelles disassemble and release drug rapidly in acidic environment that resembles lysosomal compartments. These results designated that the PTX-SPION-loaded pH-sensitive micelles were a shows potential carrier as MRI-visible drug release for colorectal cancer [6].

The nanocarrier is attached to the core and shell via a hydrogen bond, signifies an intelligent, biodegradable, and pH-responsive nanocarrier for breast cancer targeting. These drug nanocarriers were attached by hydrogen bond from -OH on mesoporous silica nanoparticle (MSN) and -NH$_2$ on chitosan (CS). When the nanocarriers exposed to the acidic tumor environment, the CS shell swells into a loose random coil, exposing the drug and building them easy to be released. The pH-sensitive delivery revealed higher drug release at pH 6.8 than at 7.4, which is suitable for breast cancer targeting [7].

### 1.2.2   REDOX

Redox potential differences are at the tissue and cellular level. For instance, the glutathione/glutathione disulfide couple has been confirmed as the most significant redox couple in animal cells, where GSH is exist at a level that is two to three orders of degree higher in the cytosol than in the extracellular fluid [8]. Moreover, studies with a rodent model have demonstrated a higher GSH concentration at tumor tissue than the normal tissues [9]. Furthermore, the reducing conditions, reactive oxygen species (ROS) are also subordinated with different pathological conditions including cancer, stroke, arteriosclerosis, and tissue injury. Disulfides convert to thiols in the existence of reducing agents, including GSH, and the resulting thiol groups can reversibly reform disulfide bonds by oxidation reaction. The mild reaction conditions of thiol-disulfide conversation also reduce it an appealing method to build disulfide-containing materials [10]. Disulfides have also been integrated into substantial systems in the form of disulfide-containing crosslinkers [11, 12]. Furthermore, diselenide linkage is another redox-responsive motif used for manufacturing redox-sensitive materials [13, 14]. In the latest study, micellar aggregates self-assembled from a diselenide-containing block copolymer exhibited high sensitivity to both oxidants and reductants [15]. Besides, the library of reduction-sensitive materials has recently been extended with the improvement of functional groups such as cis,cis,trans-diamminedichlorodihydroxy-platinum(iv) (DHP) or cis,cis,trans-diamminedichlorodisuccinato-platinum (DSP) and trimethyl-locked benzoquinone (TMBQ) [16, 17].

   The ROS mainly target by the oxidation-responsive materials such as hydrogen peroxide ($H_2O_2$) and hydroxyl radicals. Sulfur-based materials are the foremost class of oxidation-responsive material. Investigators copolymerized oxidation-convertible poly(propylene sulfide) (PPS) with polyethylene glycol (PEG) to generate amphiphiles capable of self-assembling [18]. Moreover, efficient gene delivery has been accomplished with thioketal-containing materials [19]. Ferrocene-containing materials have also been profoundly explored outstanding to the redox-sensitivity materials [20]. Additionally, initial responsive motifs such as boronic ester groups and phenylboronic acid (PBA) derivatives have also involved significant attention [21–24]. For instance, aryl boronic esters were altered at the hydroxyl groups of dextran as well as the lysine residues of RNase A for $H_2O_2$-triggered protein release and action recovery, respectively.

Recent research reported osteoarthritis targeted anti-inflammatory drug-delivery system connected with $H_2O_2$, the poly(lactic-co-glycolic) acid (PLGA) hollow microsphere carrier was incorporated with an anti-inflammatory drug, an acid precursor (consist of ethanol and $FeCl_2$), and a bubble-producing agent, sodium bicarbonate [25]. The material was intended to permit $H_2O_2$ diffusion through the microspheres allowing for ethanol oxidation to create an acidic milieu, in which sodium bicarbonate decomposed to produce $CO_2$ bubbles, distracting the shell wall of the microspheres and leading to the release of the anti-inflammatory cargo.

An observing confined ROS level has the impending to develop the diagnosis and treatment of several diseases, ranging from cardiovascular diseases to drug-induced organ failure. Integrating ROS-responsive materials signifies an appealing methodology for evolving enzyme-free ROS sensors. A lately established biosensor that uses a thin film of a hydrogel polymer containing ROS-degradable thiocarbamate linkages in its backbone was able to detecting drug-induced liver injury by observing the oxidative stress in the blood [26].

The recent research on the development of PLGA nanoparticle core and a redox-responsive amorphous organosilica shell has been successfully done. In which the outer layer is generated by self-assembly of silicate ions with a disulfide linkage containing silsesquioxane. These organic linkers act as molecular gates that can be particularly cleaved by reducing agents. This method is primarily appropriate for storage and release of hydrophobic drugs, such as docetaxel (DTX), as the treatment with reducing agents leaves open doors that permit for the discharge of DTX in the organic matrix. These methods have an improved control and slower release of the entrapped drug than bare PLGA nanoparticles, are practically stable in physiological medium and show higher cytotoxic activity over HeLa cells than the free drug [27].

Moreover, another recent research reported, a type of PEGylated core/shell structured composite nanoparticle via precipitation polymerization method. In which a disulfide-cross-linked poly(N-vinylcaprolactam-co-methacrylic acid) (P(VCL-s-s-MAA)) polymer shell was produced to proceed as sheddable thermo/pH-responsive gatekeepers, and a carboxylic acid customized mesoporous silica nanoparticles (MSN-COOH) core was suitable as an accessible reservoir to entrapped high payload. The P(VCL-s-s-MAA)-PEG shell experiences a characteristic alteration from a swollen state in pH 7.4 to a collapsed state in pH 5.0 at physiological states. On the other hand, adequately stable in water, the composite nanoparticles were sensitive to fast dissociation and rupture when exposing to 10 mM GSH, due to the shedding of polymer walls via reductive cleavage of intermediate disulfide bonds, as a result the polymer shell was active in moderating the diffusion of surrounded drugs in-and-out of MSN channels. Therefore, these stimuli-sensitive composite nanoparticles with a reductively sheddable and thermo/pH-responsive polymer shell gate might, in standard, be applied for *in vivo* cancer targeting, and synergistic drug delivery can be accomplished "just in time" in a precise incidence over the site [29].

### *1.2.3   ENZYMES*

Various roles of enzymes have in different biological processes; enzyme dysregulations associated with disease have recently developed an emergent target for medications. For illustration, ester bonds are often incorporated for targeting phosphatases, intracellular acid hydrolases, and numerous other esterases. Amides, although relatively stable to chemical attack in physiological environments, are exposed to enzymatic digestion and have been used for assembling materials sensitive to hydrolytic proteases, such as prostate-specific antigens; and materials containing cleavable azo linkers can target bacterial enzymes in the colon for site-specific drug release. Here, we elucidate some typical enzymatic triggers. In tumor invasion and metastasis, upregulation of matrix metalloproteinases (MMPs) is the major cause. The upregulated expression of MMPs inside the tumor microenvironment (TME) can assist as target-specific biological signals for triggering bioresponsive materials [30–32]. Activatable CPPs (ACPPs) with blocked cellular interaction were fabricated by combining CPPs with an anionic inhibitory domain [33, 34]. ACPPs can be stimulated by overexpressed MMPs at the tumor site through the cleavage of the linker attaching the cationic and anionic domain to visualizing tumors during surgery. Moreover, the TME, MMP upregulation is also connected with other diseases such as asthma and inflammatory bowel diseases, which also providing a possible drug-delivery target [35, 36]. In the latest research, a negatively charged hydrogel was attached to the surface of the inflamed colon with up-regulated MMP expression, where it released anti-inflammatory drugs only upon enzymatic digestion [35]. Furthermore, being a natural habitat of a sequence of bacteria that constantly secrete several enzymes including various polysaccharides. The colon is an appropriate target for site-specific drug release using polysaccharide-based materials. Biocompatible polysaccharides based polymers, such as CS, pectin, and dextran, have been used for colon-specific drug delivery via the oral route in various dosages including tablets, capsules, hydrogels, and drug conjugates carriers. Additionally, polysaccharides have also been used as cross-linkers to form lysozyme-sensitive nanogels, which were further fused into contact lenses for sustained release of glaucoma drugs [37]. The added class of enzymes whose expression is often raised at the tumor site is hyaluronidase (HAase). In a gel-based liposome nanoformulation [38], the core-shell designed nanocarrier was intended to have a cell-penetrating peptide-modified liposome core for chemotherapeutic drug loading and a hyaluronic acid (HA)-based cross-linked shell for the entrapment of tumor necrosis factor-related apoptosis-inducing ligand (TRAIL), which is a cytokine that can be attached to the death receptors on the plasma membrane. The HA shell was digested by the overexpressed HAase at TME, leading to the release of TRAIL, followed by the sequential release of DOX inside the tumor cells. Besides, to activating cargo release, the HAase has also been used for the *in situ* assembly of extracellular drug-delivery depots for sustained confined cargo release [39]. Furthermore, intracellular enzyme protein kinase Cα (PKCα) is essential for cancer cell proliferation [40] that hyper-activated in various cancer cell lines but displays low activity in normal cells. To take benefit of the cancer-specificity of PKCα, investigators have fabricated polymers equipped with PKCα-specific peptide substrates for targeted gene delivery. In one more study, caspase 3-cleavable polymeric nanocarriers that were cross-linked by the caspase 3 substrate-based peptide cross-linker were established to transport caspase 3 for stimulating apoptosis of cancer cells in a 'self-degradable' manner [41]. A number of phospholipases also function in the extracellular microenvironment when designated as therapeutic targets [42]. In a hydrogel-based closed-loop system intended to be responsive to blood coagulation, heparin release was activated by the cleavage of thrombin-degradable peptide when the blood thrombin reached a high level; the released heparin then deactivated thrombin to inhibit further drug release [43].

A member of the pro-protein convertase family, furin has a key role in tumor development, metastasis, and angiogenesis. In one investigation, a furin-cleavable peptide cross-linker was fused with drug-delivery carriers, which could be degraded gradually to release payload protein along their cellular uptake pathway [44]. In a recent investigation, a graphene-based co-delivery system took benefit of the trans-membrane activity of furin to cleave the furin-degradable substrate for the exposure of TRAIL in the direction of the membrane [45].

The recent research developed on a facile strategy to synthesize HA conjugated mesoporous silica nanoparticles (MSPs) for targeted enzyme responsive drug delivery. The system anchored HA polysaccharides not only act as capping agents but also as targeting moiety without the need of additional alteration. The nanoconjugates acquire a lot of attractive characteristics include chemical simplicity, high colloidal stability, good biocompatibility, cell-targeting ability, and precise load release, building them promising agents for biomedical applications. As a proof-of-conception expression, the nanoconjugates are shown to release payload from the interior pores of MSPs upon HA degradation in response to hyaluronidase-1 (Hyal-1). Furthermore, after receptor-mediated endocytosis into cancer cells, the anchored HA was degraded into small fragments, facilitating the release of drugs to kill the cancer cells [46].

Moreover, bioresponsive polymer structural designs can authorize medical therapies by appealing molecular feedback-response mechanisms resembling the homeostatic alteration of living tissues to varying environmental restrictions. These blood coagulation-responsive hydrogel systems can deliver heparin in amounts activated by the environmental levels of thrombin, the main enzyme of the coagulation cascade, which—in sequence—becomes inactivated due to released heparin. The bio-responsive hydrogel quantitatively reduces blood coagulation over some hours in the occurrence of pro-coagulant stimuli and during repeated incubation with fresh, non-anticoagulated blood. These characteristics facilitate the introduced material to offer sustainable, autoregulated anticoagulation, addressing the main challenge of numerous medical therapies. The explored concept may assist the development of materials that permit the efficient and controlled application of drugs and biomolecules [47].

Furthermore, the improvement of a bioresponsive drug delivery system needs exquisite engineering of materials so that they are capable to respond to the signals stemming from the physiological environment. In the present investigation, a novel Pluronic® based thermogelling system comprising matrix metalloproteinase-2 (MMP2) responsive peptide sequences. A novel thermosensitive multiblock copolymer consist of an MMP2-labile octapeptide (Gly-Pro-Val-Gly-Leu-Ile-Gly-Lys) was synthesized from Pluronic® triblock copolymer. The polymer was intended to form thermogel at body temperature and degrade in the occurrence of MMP overexpressed in the tumor. The synthesized polymer was a multiblock copolymer with ~2.5 units of Pluronic®, which was in solutions, exhibited a reverse thermal gelation around body temperature. The gelation temperatures of the multiblock copolymer solutions were lower than those of the equivalent Pluronic® monomer at a particular concentration. The cytotoxicity of the synthesized polymer was lesser compared to its monomer. The solubility hydrophobic drug, paclitaxel was enhanced in the polymer solutions through micelle formation. The synthesized polymer was specifically degraded in the presence of MMP. Paclitaxel release was dependent on the enzyme concentration.

### 1.2.4   GLUCOSE

In recent days, the blood sugar monitoring and open-loop intelligent drug delivery based insulin injection treatment is still the crucial management of type 1 and advanced type 2 diabetes [49]. Apart from existence both painful and inconvenient, it is enormously challenging to tightly regulate blood glucose levels (BGLs) as follows, which clues to a great threat of diabetes complications. Likewise, hypoglycaemias can consequence in fatal insulin shock. Hence, there is a remarkable need for glucose-responsive 'closed-loop' treatments that mimic the function of the healthy pancreas and work in a self-regulated manner.

The principal glucose-responsive insulin delivery system (GRIDS) was designed in 1979 using concanavalin A (Con A), a member of the saccha-ride-binding lectin family. Free glucose incorporated within the specific binding sites of the Con A-polymer complex, triggering the dissociation of the complex and consequent insulin release [50]. Huge efforts have been devoted to this field [49]; concentrating on achieving a fast response, ease of administration and outstanding biocompatibility. Normally, there are two kinds of GRIDS established on the mechanism of regulating the BGL. The first type, such as GRIDS based on Con A or synthetic boronic acids (BA) [51–53], a high BGL assists a straight activation of insulin release (Directly triggered model). The second type systems, which relate the glucose oxidase (GOx)-based enzymatic reaction, a high BGL brings a fall of local pH or oxygen level, which successively stimulates insulin release (Progressively activated model) [54].

The saccharide-sensitivity system developed by the well-established BA-diol interaction concentrates BA-comprising polymers impending candidates for assembling glucose-responsive polymer. PBA with electron-withdrawing groups is commonly used as glucose-responsive material [52, 55]. In the latest investigation, scientists linked small molecules consist of both an aliphatic and a PBA moiety to insulin. Such conjugation afforded binding to serum albumin, or other hydrophobic constituents in serum, for sustained circulation half-life as well as the glucose-responsive release of insulin [55]. In an additional instance, polymersomes assembled from polyboroxole block copolymers revealed on demand insulin release in the presence of high BGL [56].

The glucose-responsive swelling and deswelling of GOx-immobilized hydrogels fabricated with cationic copolymers [53, 63]. In this system, GOx transforms glucose into gluconic acid in the existence of oxygen, resultant in a declined local pH value [53, 57]. This mechanism can increase the solubility of lysine-modified insulin [57], which activates the swelling or collapsing of hydrogels [53, 58–61] or the dissociation of nanoformulations [62], leading to insulin release. Furthermore, about the localized acidification, investigators have recently reserved improvement of enzymatically produced local hypoxia (prompted by high BGL) to build glucose-responsive microneedle-array patches (also called smart insulin patches) [64]. Rapid glucose-responsive insulin release can be accomplished through vesicles accumulated from 2-nitroimidazole-conjugated HA.

The glucose-responsive materials are not used for GRIDS. For illustration, in a current study, PBA derivatives based hydrogel formulations showed a potential role as both the delivery carrier for protein therapeutics and substrates for 3D cell culture [65]. The hydrogel demonstrated self-healing (quick structural recovery) based on the active interaction between PBA and diols. A nanostructured surface created by attaching a PBA-containing brush from an aligned silicon nanowire array captured and released cells in response to alterations in pH values and glucose concentration [66]. Moreover, the recent investigation of microneedle-based cancer immunotherapeutic delivery system targeting melanoma, in which GOx transformed blood glucose to localized acidification for activating the sustained release of anti-PD1 antibody with improved retention in the tumor site [67].

Besides, to drug-delivery systems, glucose-responsive constituents are also used for longstanding glucose monitoring. For instance, single-walled carbon nanotubes (SWNTs) functionalized with a glucose analog was established for glucose-responsive [68, 69]. The SWNTs forms accumulations in the existence of saccharide-binding Con A or PBA, extinguishing the fluorescence indication. By contrast, the dissociation of such aggregates due to the competitive binding of glucose clues to the retrieval of the fluorescence. Fluorescent polyacrylamide hydrogel beads invented from monomers containing glucose-recognition sites and fluorogenic sites have exposed the potential for continuous BGL monitoring [70].

There are several mechanisms by which glucose-sensing triggers can be integrated with nanoparticle design to facilitate glucose-responsive behavior. Nanoparticles prepared using polymers that are molecularly imprinted with glucose and phenylboronic acid (PBA) could form supramolecular assemblies through reversible hydrogen-bonding interactions between glucose and PBA molecules. These nanoparticle assemblies would then be sensitive to glucose concentrations in their localized environment through the competitive binding of glucose from the environment to PBA. Alternatively, glucose-imprinted polymers could be combined with glucose-binding proteins such as concanavalin A (Con A) to form supramolecular assemblies that are similarly responsive to glucose. Glucose-sensitive nanoparticle systems can also be engineered by combining pH-sensitive polymers with the glucose-sensitive enzyme GOx, which enzymatically converts glucose to gluconic acid, producing a drop in pH in the nanoparticle microenvironment. The triggers can be integrated within a nanoparticle that is engineered to disassemble by either swelling or degrading in response to increased glucose levels, thus providing a mechanism by which the insulin cargo can be released and made bioavailable [71].

The design of drug delivery nanocarriers having targeted recognition followed by bioresponsive-controlled release, especially via glucose-responsive release, is a challenging issue. Here, we report magnetic mesoporous silica (MMS)-based drug delivery nanocarrier that can target specific cells and release drugs via a glucose-responsive gate. The design involves the synthesis of MMS functionalized with phenylboronic acid and folate. After drug loading inside the pores of MMS, outside of the pores are closed by dextran via binding with phenylboronic acid. Dextran-gated pores are opened for drug release in the presence of glucose that competes binding with phenylboronic acid. We found that tolbutamide and camptothecin loaded MMS can target beta cells and cancer cells, respectively, release drugs depending on bulk glucose concentration and offers glucose concentration-dependent cytotoxicity. Developed functional MMS can be used for advanced drug delivery applications for diabetes and cancers with more efficient therapy [72].

Glucose-sensitive drug delivery systems, which can continuously and automatically regulate drug release based on the concentration of glucose, have attracted much interest in recent years. Self-regulated drug delivery platforms have potential applications in diabetes treatment to reduce the intervention and improve the quality of life for patients. At present, there are three types of glucose-sensitive drug delivery systems based on GOx, Con A, and phenylboronic acid (PBA), respectively. This review covers the recent advances in GOx-, Con A-, or PBA-mediated glucose-sensitive nanoscale drug delivery systems, and provides their major challenges and opportunities [73].

In recent years, glucose-sensitive drug delivery systems have attracted considerable attention in the treatment of diabetes. These systems can regulate payload release by the changes of BGLs continuously and automatically with potential applications in self-regulated drug delivery. BA, especially phenylboronic acid (PBA), as a glucose-sensitive agent has been the focus of research in the design of glucose-sensitive platforms. This article reviews the previous attempts at the developments of PBA-based glucose-sensitive drug delivery systems regarding the PBA-functionalized materials and glucose-triggered drug delivery. The obstacles and potential developments of glucose-sensitive drug delivery systems based on PBA for diabetes treatment in the future are also described. The PBA-functionalized platforms that regulate drug delivery induced by glucose are expected to contribute significantly to the design and development of advanced intelligent self-regulated drug delivery systems for the treatment of diabetes [74].

### 1.2.5   IONS

All biological fluids have varied Ionic strength. For instance, each gastro-intestinal site has a specific ionic concentration; hence, polymer sensitive to ionic strength are of specific interest for oral delivery carriers [75]. Furthermore, gradients in ionic concentration also occur in the blood, and in interstitial and intracellular compartments, corresponding to other drug administration methods, such as intravenous injection.

A major family of physical ion-responsive materials is ion-exchange resins, which are often used for different purposes such as to mask the taste of the bitter drug, as a counter ion-responsive drug release carrier, and to achieve sustained drug release [76]. These resins are typically insoluble polymers consists of a cross-linked polystyrene backbone with side chains containing ion-active groups such as sulfonic acid and carboxylic acid. The counter ions present in the saliva and gastrointestinal fluids stimulate drug release, which is directed by an equilibrium exchange reaction via oral administration. For illustration, cationic polymers holding quaternary ammonium groups show sensitivity towards ions in the saliva [77]. Polymers displaying a lower critical solution temperature (LCST) also show definite sensitivity towards ionic strength [78]. The LCST can be moved, generally to a lower temperature, in the existence of salt, following the Hofmeister series [78]. Polyion complex micelles signify another foremost family of ionic strength-sensitive materials. Reversible development and dissociation of polyion complex micelles over an alteration in salt concentration (and thus ionic strength) have been used for the release of cargo in a controlled manner [79]. In addition to responding physically to variations of ionic strength, materials can also react to specific ion types, normally by establishing complexes. In the recently reported investigation, metal-ion-responsive adhesive hydrogel, modified with β-cyclodextrin and hydrophobic 2,2′-bipyridyl moieties, the chemically selective adhesion property could be changed by governing the inclusion of inhibitory metal ligands to host moieties [80].

Some kinds of cations and anions are contained in body fluids such as blood, interstitial fluid, gastrointestinal juice, and tears at relatively high concentrations. Ion-responsive drug delivery is therefore available to design the unique dosage formulations which provide optimized drug therapy with effective, safe, and convenient dosing of drugs. It uses the change of ionic concentration as a trigger to perform functions after dosing, such as controlled drug release, site-specific drug release, *in situ* gelations, prolonged retention at the target site, and enhancement of drug permeation. Ion-exchange resins/ fibers, anionic or cationic polymers, polymers exhibiting transition at LCST, self-assemble supramolecular systems, peptides, and metal-organic frameworks have been researched and reported to respond to the body fluid ions. Administration of ion-sensitive formulations via oral, ophthalmic, transdermal, and nasal routes showed many advantages. In this review, recent studies on ion-responsive drug delivery systems are summarized, and the progress and prospect of this field are discussed [81].

Functional and smart polymers are currently playing a significant role in formulations of controlled drug delivery systems due to their responsive behavior towards environmental stimuli. The delivery systems must be non-toxic, non-immunogenic, and must be having optimum trapping and release properties for an active agent. To be more effective and economical the controlled devices should be able to release the active agent in a control and site-specific manner. The physical and chemical properties of the polymers provide opportunities to design therapeutic devices for various applications. Though various delivery systems based on electrical, mechanical, and viral systems have been fabricated with great successes but these delivery systems have shown poor transfection efficiency and found to be immunogenic; hence, delivery systems based on functional polymers such as poly(ethylene glycol) and dendric poly(amidoamine) found to be of great significance. The delivery systems using natural and biodegradable polymers such as CS, pectin, and polysaccharides proved to be more are acceptable due to their biocompatibility and biodegradability in physiological fluids in comparison to synthetic polymer systems. Considering the importance of natural polymers, the pH and ion responsive drug delivery systems have been designed using different forms and derivatives of the CS. These delivery systems have been tested for efficiency of loading and delivery of active agents as a function of solution pH and ionic strength of the medium. The naturally occurring CS has provided enormous opportunities for controlling its properties to fabricate control site-specific delivery systems. The degree of deacetylation in CS proved to be significant in controlling its stimuli-responsive properties for drug delivery systems. The nano-sized CS delivery systems found to be more therapeutic in comparison to macro and micro-sized delivery devices for controlled and sustained delivery of the active agents. The role of various parameters would be discussed and highlighted in this talk [82].

### 1.2.6   ATP

Adenosine-5'-triphosphate (ATP) often referred to as the 'molecular unit of currency' of intracellular energy transfer, which creates a higher concentration intracellularly than in the extracellular environment because of its instant relationship with cell metabolism. The ATP-controlled drug-delivery approach frequently uses ATP-targeted aptamers as 'biogates' to accomplish on demand payload release [83]. Nowadays, many formulations, such as mesoporous silica-based drug carrier, polyion micelles, aptamer-cross-linked DNA microcapsules, tubular structures assembled from proteins and nanogels composed of DNA complexes, have established the capability to release drug or reinstate fluorescence signals on experience to relatively concentrated intracellular ATP [84–87]. In the present case, ATP either competitively binds to drug loading sites to activate payload release or fuels conformational variations, which produce structure-disrupting forces. For instance, DNA aptamers were fused into a Dox-loaded DNA duplex [86]. The competitive binding of ATP molecules to ATP aptamers in an ATP-rich atmosphere resulted in the dissociation of the duplex for targeted drug release. In additional strategy [85], a protein nanotube assembled from barrel-shaped chaperonin units protected payload molecules from biological degradation, but upon exposure to the intracellular hydrolysis of ATP, the tempted conformational conversion of the chaperonins led to the disassembly of the tubular structure and the release of the therapeutic molecules. In place of working exclusively, ATP can assist as part of a combinational activate with other stimuli [88].

In a polymeric nanogel based nanocarrier [5], the ssDNA aptamer was hybridized with its complementary nucleotides to form a DNA duplex, which contains a "GC" pair for loading the small molecule anticancer drug DOX. In the absence of ATP, the DNA duplex was rather stable to hold the DOX payload; while high concentrations of ATP will compete with the complementary DNA to bind the ATP aptamer, dissociating the duplex and releasing the loaded drug. To neutralize the strong negative charge of the DNA duplex, thus condensing its size in the solution for packing into a nanoparticle, the positively charge peptide protamine was complexed with the DNA duplex. The peptide/DNA complex loaded with DOX formed the core of the particle, upon which a negatively charged polymer HA was coated. The HA coating served three main purposes: (1) it protected the peptide/DNA core from premature degradation while in circulation; (2) the HA is an active targeting ligand that binds receptor-like CD44 and RHAMM on cancer cell membrane; and (3) the enzyme HAase overexpressed in TME could function as an extracellular trigger in addition to intracellular ATP for prompting drug release. In an *in vivo* xenograft tumor model, systemic administration of the nanogel based particle exhibited higher accumulation at the tumor (~4 fold more DOX) owing to passive as well as active targeting effects as compared with non-gel coated particles. After exposing the peptide/DNA duplex per HAase degradation, the positively charged protamine stimulated endosome escape, escorting the DNA duplex loaded with DOX into the cytosol where the high levels of ATP triggered DOX release. The released DOX gradually accumulated in the nucleus by diffusion. Formulations with the ATP aptamer showed significantly higher tumor growth inhibition effect than non-ATP responsive control groups [89].

### 1.2.7  HYPOXIA

Various diseases such as cancer, cardiomyopathy, ischemia, rheumatoid arthritis, and vascular diseases are normally associated with hypoxia [90]. Tumor hypoxia is usually measured as a negative prognostic as of its essential role in tumor progression and therapy resistance, and has been widely exploited for manufacturing diagnostic agents and therapeutics [91]. Nitroaromatic derivatives that can be converted to hydrophilic -aminoimidazoles under hypoxic circumstances with a relatively high sensitivity are amongst the most extensively exploited functional motifs for hypoxia imaging and the design of bioreductive prodrugs. Likewise, azobenzene (AZO), alternative firm, hypoxia-sensitive motif formerly used as an imaging probe, has been fused in the form of a bioreductive linker for targeted siRNA delivery [92]. In recent times, investigators also used the knowledge of engineering smart delivery systems comprising oxygen-sensitive groups as alternatives for hypoxia-sensitive small molecules to increase the sensitivity and specificity for *in vivo* imaging [93, 94]. For instance, ultrasensitive detection of cancer cells, a water-soluble macromolecular imaging probe with hypoxia-sensitivity and near-infrared (NIR) emission was manufactured by conjugating a phosphorescent iridium (iii) complex to a hydrophilic polymer, poly(N-vinylpyrrolidone) (PVP) [94]. Related methods might be of interest for improving imaging for the diagnosis and real-time observings of other hypoxia-associated diseases, such as stroke and ischemia [95, 96].

Hypoxia in tumors contributes to overall tumor progression by assisting in epithelial-to-mesenchymal transition, angiogenesis, and metastasis of cancer. In this study, we have synthesized a hypoxia-responsive, diblock copolymer poly(lactic acid)-azobenzene-poly(ethylene glycol), which self-assembles to form polymersomes in an aqueous medium. The polymersomes did not release any encapsulated contents for 50 min under normoxic conditions. However, under hypoxia, 90% of the encapsulated dye was released in 50 min. The polymersomes encapsulated the combination of anticancer drugs gemcitabine and erlotinib with entrapment efficiency of 40% and 28%, respectively. We used three-dimensional spheroid cultures of pancreatic cancer cells BxPC-3 to demonstrate the hypoxia-mediated release of the drugs from the polymersomes. The vesicles were nontoxic. However, a significant decrease in cell viability was observed in hypoxic spheroidal cultures of BxPC-3 cells in the presence of drug encapsulated polymersomes. These polymersomes have the potential for future applications in imaging and treatment of hypoxic tumors [97].

Hypoxia is a condition found in various intractable diseases. Here, we report self-assembled nanoparticles which can selectively release the hydrophobic agents under hypoxic conditions. For the preparation of hypoxia-responsive nanoparticles (HR-NPs), a hydrophobically modified 2-nitroimidazole derivative was conjugated to the backbone of the carboxy-methyl dextran (CM-Dex). DOX, a model drug, was effectively encapsu-lated into the HR-NPs. The HR-NPs released DOX in a sustained manner under the normoxic condition (physiological condition), whereas the drug release rate remarkably increased under the hypoxic condition. From *in vitro* cytotoxicity tests, it was found the DOX-loaded HR-NPs showed higher toxicity to hypoxic cells than to normoxic cells. Microscopic obser-vation showed that the HR-NPs could effectively deliver DOX into SCC7 cells under hypoxic conditions. *In vivo* biodistribution study demonstrated that HR-NPs were selectively accumulated at the hypoxic tumor tissues. As a consequence, drug-loaded HR-NPs exhibited high anti-tumor activity *in vivo*. Overall, the HR-NPs might have a potential as nanocarriers for drug delivery to treat hypoxia-associated diseases [98].

## 1.2.8 TEMPERATURE

Smart polymers that show a LCST usually experience rapid phase transi-tion nearby the LCST, which can be simply regulated by altering the ratio of hydrophobic and hydrophilic components or by substituting the end groups [99]. The polymer is capable with intrinsic sensitivity to physi-ological temperature; when the LCST of a polymer is existed between room temperature and body temperature. Sol-gel transition at its LCST of 32°C temperature can be further optimized to be nearer to body temperature through copolymerization with hydrophobic monomers or the addition of hydrophobic groups [100]. The use of poly(N-isopropyl acrylamide (PNIPAM) as a thermoresponsive drug carrier was established in the 1980s [101, 102]. In certain fusion systems, the LCST of PNIPAM-based materials were altered by integrating inorganic materials, such as gold nanoparticles [103–105]. To speed up the thermoresponsive transition course, scientists have established controllable, triggered nanogels as cross linkers for the construction of thermoresponsive hydrogels, which displayed quick and reversible responsive features although preserving high elasticity [106]. Homopolymers and copolymers of N-acryloyl pyrrolidine (APy) and 2-hydroxyethyl methacrylate (HaEMA) have also been produced to accomplish temperature-regulated insulin release by the modification of permeability [107, 108]. Besides, the temperature-sensitive coiled-coil domains of proteins have been complexed onto soluble polymers to create thermoresponsive polymer [109].

### 1.2.9   MECHANICAL CUES

The narrowing or blockade of blood vessels marks in a substantial deviation in fluid shear force between healthy and confined blood vessels. To target diseased blood vessels with blockade, by the abnormally high shear stress at the blockage site as an attractive newly developed approach. A wearable, tensile-strain-sensitive device composed of stretchable elastomer film with inserted drug-entrapped PLGA microspheres was lately developed [110]. On use of strain, the mircoparticles experienced surface widening and compression, thus releasing the drug.

### 1.3   APPLICATIONS OF BIO-RESPONSIVE POLYMERS

Bio-responsive polymers are not just for drug delivery. Their properties make them especially suited for bio-separations. The time and costs involved in purifying proteins might be reduced significantly by using Bio-responsive polymers that undergo rapid reversible changes in response to a change in medium properties. Conjugated systems have been used for many years in physical and affinity separations and immunoassays.

Microscopic changes in the polymer structure are manifested as precipitate formation, which may be used to aid separation of trapped proteins from solution.

These systems work when a protein or other molecule that is to be separated from a mix, forms a bio-conjugate with the polymer, and precipitates with the polymer when its environment undergoes a change. The precipitate is removed from the media, thus separating the desired component of the conjugate from the rest of the mixture. Removal of this component from the conjugate depends on recovery of the polymer and a return to its original state, thus hydrogels are very useful for such processes [111].

Another approach to controlling biological reactions using Bio-responsive polymers is to prepare recombinant proteins with built-in polymer binding sites close to ligand or cell-binding sites. This technique has been used to control ligand and cell-binding activity, based on a variety of triggers including temperature and light.

Bio-responsive polymers play an essential part in the technology of self-adaptive wound dressings. The dressing design presents proprietary super-absorbent synthetic Bio-responsive polymers immobilized in the 3-dimensional fiber matrix with added hydration functionality achieved by embedding hydrogel into the core of the material [112].

The dressing's mode of action relies on the ability of the polymers to sense and adapt to the changing humidity and fluid content in all areas of the wound simultaneously and to automatically and reversibly switch from absorption to hydration. The smart polymer action ensures the active synchronized response of the dressing material to changes in and around the wound to support the optimal moist healing environment at all times [113].

Modern bioresponsive materials are gradually developed to interface with biological tissues in well-defined ways. Significant classes of bioresponsive materials are those that are highly hydrated-hydrogels. Bioresponsive hydrogel polymers, which have been developed to engaged in a conversation with their biological surroundings. A biological incident takes place upon interaction with the material, which can be accomplished by introducing target moiety that binds specific biomolecules into the material structure. These target moieties offer instructions to manage or direct biological interactions. Hydrogels made up of elastic cross-linked networks with interstitial spaces that include as much as 90–99% w/w water. Hydrogels are prepared by different methods such as chemical polymerization or by physical self-assembly of synthetic or naturally occurring building blocks.

Bioresponsive polymers present a fascinating alternative for the delivery of genes and other therapeutic nucleic acids. In the present delivery process, the polymeric carriers face lots of diverse delivery tasks and unusual physiological microenvironments. Polymers can be developed to respond to microenvironmental variations with alters in their physiochemical properties, enabling them to execute individual delivery tasks. Breakage of covalent bonds, disassembly of noncovalent interactions, alters of protonation, conformation, or hydrophilicity/lipophilicity, can activate such dynamic physicochemical alterations. The polymeric transporters have to stably attached with therapeutic nucleic acid during the extracellular delivery phase and defend it against degradation in the bloodstream. At the intracellular site of action, the polyplex has to disassemble to a level that the nucleic acid is functionally available. Polyplexes require to be protected in the blood circulation and be inert against various potential biological interactions, but should actively interact with the target cell surface by electrostatic or ligand-receptor interactions. Lipid-membrane deterioration at the cell membrane or nontarget sites is typically connected with undesired cytotoxicity, the similar biophysical occasion, though, is needed within an endocytic vesicle for polyplex transfer into the cytosol. Bioresponsive polymers can be developed and integrated into polyplexes. For instance, dynamic stabilization of the polymer/nucleic acid core and transient establishment of properties needed for crossing lipid-membrane barriers. Bioresponsive delivery fields at the polyplex surface involved for shielding, deshielding, and cell targeting also put into better performance.

The novel use of reducible hyperbranched (rHB) polymers for delivery of RNA interference (RNAi) therapeutics. Cationic poly(amidoamine) hyperbranched polymers that enclose different contents of reducible disulfide to nonreducible linkages (0%, 17%, 25%, and 50%) were employed to form inter polyelectrolyte polyplexes with small interfering RNA (siRNA) and precursor microRNA (miRNA). The rHB complexes of ~100 nm in size, which revealed redox-activated disassembly in the presence of dithiothreitol (DTT). The complexes were keenly internalized and showed no cellular toxicity in an endogenous improved green fluorescence protein (EGFP) expressing H1299 human lung cancer cell line. The highest specific enhanced green fluorescent protein (EGFP) gene silencing (~75%) was accomplished with rHB (17%)/siRNA complexes at a weight-to-weight (w/w) ratio of 40 that associated with the capability for this polymer to effectively transfect pre-miRNA. The role of particle disassembly for intracellular targeting and modulation of gene silencing addressed in this work are significant considerations in the development of this and other next-generation delivery systems.

Gene therapy creates a center of attention because of its potential for therapeutic interventions in cancer and other diseases. Nonviral gene vectors have been used for more than three decades. The development of novel carriers with decreased cytotoxicity, improved delivery, and enhanced transfection efficiency has remained an important aim for nonviral gene delivery. By structural modifications such as conformational alters or chemical bond cleavage, bioresponsive polycations can enhance the transfection properties of polyplexes—formed by cationic polymers to deliver nucleic acids into cells—at different steps of delivery. These polycations can react to subtle differences between the extracellular and intracellular surroundings. Beside with noncovalently formed polyplexes, the progress of covalent polymer-oligonucleotide conjugates with bioresponsive linkers is promising as nucleic acid therapeutics. Numerous novel cationic polymers with bioresponsive and admirable biocompatible properties have newly been developed.

A stimuli-sensitive or smart polymer undergoes an abrupt change in its physical properties in response to a small environmental stimulus. These polymers are also called intelligent polymers because small changes occur in response to an external trigger until a critical point is reached, and they have the ability to return to their original shape after the trigger is removed. The exclusivity of these polymers lies in their nonlinear response triggered by a very small stimulus and which produces noticeable macroscopic alterations in their structure. Various stimuli are responsible for controlling drug release from smart polymeric drug delivery systems. These transitions are reversible and include changes in physical state, shape, and solubility, solvent interactions, hydrophilic, and lipophilic balances and conductivity [118–120]. The driving forces behind these transitions include neutralization of charged groups by the addition of oppositely charged polymers or by pH shift, and change in the hydrophilic/lipophilic balance or changes in hydrogen bonding due to increase or decrease in temperature. The major benefits of smart polymer-based drug delivery systems include reduced dosing frequency, ease of preparation, maintenance of desired therapeutic concentration with a single dose, prolonged release of the incorporated drug, reduced side effects and improved stability [121, 122].

Thermosensitive polymers undergo an abrupt change in their solubility in response to a small change in temperature. An aqueous thermosensitive polymeric solution exhibits temperature-dependent and reversible sol-gel transitions near body temperature that control the rate of release of the incorporated drug along with maintaining physicochemical stability and biological activity. This phenomenon is generally governed by the ratio of hydrophilic to lipophilic moieties on the polymer chain and is an energy-driven phenomenon which depends on the free energy of mixing or the enthalpy or entropy of the system. A common characteristic feature of thermosensitive polymers is the presence of a hydrophobic group, such as methyl, ethyl, and propyl groups. These polymers possess two additional critical parameters, i.e., LCST and upper critical solution temperature (UCST) [123–126]. LCST is the temperature above which the polymeric monophasic system becomes hydrophobic and insoluble, leading to phase separation, whereas below the LCST the polymers are soluble. For polymers having LCST, a small increase in temperature results in negative free energy of the system ($\Delta G$) leading to a higher entropy term ($\Delta S$) with respect to increase in the enthalpy term ($\Delta H$) in the thermodynamic relation $\Delta G = \Delta H - T\Delta S$. The entropy increases due to water-water associations. In contrast to UCST systems, an LCST system is mostly preferred for drug delivery technologies due to the need for high temperatures for UCST systems, which is unfavorable for heat-labile drugs and biomolecules. According to the phase response to the temperature change, polymers are subdivided into negatively thermosensitive, positively thermosensitive, and thermoreversible types. Examples of conformational change that take place at the critical solution temperature are polymeric micelle packing and coil-to-helix transitions. The most commonly used LCST thermosensitive polymers include poly(*N*-isopropyl acrylamide), poly(*N,N*-diethylacrylamide), poly(*N*-vinylalkylamide), poly(*N*-vinylcaprolactam), phosphazene derivatives, pluronics, tetronics, polysaccharide derivatives, CS, and PLGA-PEG-PLGA triblock copolymers [127].

Poly(*N*-isopropyl acrylamide) is a thermosensitive polymer that exhibits a sharp LCST at 32°C that can be shifted to body temperature by formulating with surfactants or additives. These polymers exhibit unique characteristics with respect to the sharpness of their almost discontinuous transition. This makes poly(NIPAAM) an excellent carrier for *in situ* drug delivery. Gelation of 5% polymer solutions occurs at various temperatures in phosphate-buffered saline (PBS). As the temperature is increased to 27°C, the clear polymer solution became cloudy and upon further heating, the polymer solution forms a gel. At the gel-shrinking temperature of 45°C, i.e., the expulsion of water from the gel occurs. No hysteresis occurs between sol-gel and gel-sol; it reverts to the sol state upon cooling to room temperature. The use of poly NIPAAM is limited due to cytotoxicity attributed to the presence of quaternary ammonium in its structure, its non-biodegradability, and its ability to activate platelets upon contact with body fluids. Many attempts have been made to reduce the initial burst drug release associated with thermosensitive systems due to slow *in vivo* sol-gel transition. Studies proved that significant improvement in release characteristics can be achieved by optimizing the chain-length ratio between hydrophilic and hydrophobic segments. A novel triblock polymeric system PCL-PEG-PCL showed a marked reduction in initial burst release by coupling to a peptide and *in vitro* drug release studies showed a sound sustained-release profile for over one month. The major advantage of thermosensitive polymeric systems is the avoidance of toxic organic solvents, the ability to deliver both hydrophilic and lipophilic drugs, reduced systemic side effects, site-specific drug delivery, and sustained release properties [128, 129]. All pH-sensitive polymers consist of pendant acidic or basic groups that can either accept or release a proton in response to changes in environmental pH. Polymers with a large number of ionizable groups are known as polyelectrolytes. Polyelectrolytes are classified into two types: weak polyacids and weak polybases. Weak polyacids accept protons at low pH and release protons at neutral and high pH [126]. Poly(acrylic acid) (PAA) and poly(methacrylic acid) (PMAAc) are commonly used pH-responsive polyacids [130, 131]. As the environmental pH changes, the pendant acidic group undergoes ionization at a specific pH called p$K$a. This rapid change in a net charge of the attached group causes an alteration in the molecular structure of the polymeric chain. This transition to expanded state is mediated by the osmotic pressure exerted by mobile counter ions neutralized by network charges. pH-sensitive polymers containing a sulphonamide group are another example of polyacid polymers. These polymers have p$K$a values in the range of 3–11 and the hydrogen atom of the amide nitrogen is readily ionized to form polyacids. Narrow pH range and good sensitivity is the major advantage of these polymers over carboxylic acid-based polymers.

CS is a polycationic biopolymer soluble in acidic solution and undergoes phase separation at a pH range close to neutrality through deprotonation of the primary amino group by inorganic ions. The gelation mechanism of

CS occurs through the following interactions which involve electrostatic attraction between the ammonium group of the CS and an inorganic ion, hydrogen bonding between the CS chains, and CS-CS hydrophobic interactions. However, the formed gel is in further need of cross-linking agents to produce a gel with sufficient mechanical stability and to release the low molecular weight drug in a controlled manner. Several studies reported that the structural strength of CS depends on the porosity of the CS gel which in turn is a function of the crystallinity of the polymer. The structural strength of the polymer can be improved either by blending with the polymers or by hydrophobic modification of the polymer. One example includes the cross-linking of CS-polyvinylpyrrolidone with glutaraldehyde to form a semi-interpenetrating polymeric network that gels *in situ* at physiological pH.

Polybases bearing an attached amino group are the most representative polybasic group. Poly($N$, $N$-dimethylamino-ethyl methacrylate) (PDMAEMA) and poly($N$, $N$-diethylamino-ethyl methacrylate) (PDEAEMA) have been the most frequently used pH-responsive polymeric bases. The amino group is protonated at high pH and positively neutralized and ionized at low pH. PDEAEMA has a hyper coiled conformation because of the presence of longer hydrophobic groups such as ethyl groups, which induce stronger hydrophobic interactions as the aggregation force. Introducing a more hydrophobic moiety can offer a more compact conformation and a more discontinuous phase [132].

Glucose responsive polymers have the ability to mimic normal endogenous insulin secretion, which minimizes diabetic complications and can release the bioactive compound in a controlled manner. These are sugar-sensitive and show variability in response to the presence of glucose. These polymers have garnered considerable attention because of their application in both glucose-sensing and insulin-delivery applications. In spite of these advantages, the major limitations are its short response time and possible non-biocompatibility. Glucose-responsive polymeric-based systems have been developed based on the following approaches: enzymatic oxidation of glucose by GOx, and binding of glucose with lectin or reversible covalent bond formation with phenyl boronic acid moieties.

Glucose sensitivity occurs by the response of the polymer toward the by-products that result from the enzymatic oxidation of glucose. GOx oxidizes glucose resulting in the formation of gluconic acid and $H_2O_2$. For example, in the case of PAA conjugated with the GOx system, as the blood glucose level is increased glucose is converted into gluconic acid which causes the reduction of pH and protonation of PAA carboxylate moieties, facilitating the release of insulin. This system is increasingly successful due to its release pattern mimicking that of the endogenous release of insulin [133, 134].

Another system utilizes the unique carbohydrate-binding properties of lectin for the fabrication of a glucose-sensitive system. Lectins are multivalent proteins and numerous glucose-responsive materials are obtained from this glucose-biding property of lectins. The response of these systems was specific for glucose and mannose, while other sugars caused no response. Con A is a lectin possessing four binding sites and has been used frequently in insulin-modulated drug delivery. In this type of system, the insulin moiety is chemically modified by introducing a functional group (or glucose molecule) and then attached to a carrier or support through specific interactions which can only be interrupted by the glucose itself. The glycosylated insulin-Con A complex exploits the competitive binding behavior of Con A with glucose and glycosylated insulin. The free glucose molecule causes the displacement of glycosylated Con A-insulin conjugates within the surrounding tissues and are bioactive. Additional studies reported the synthesis of monosubstituted conjugates of glucosyl-terminal PEG and insulin. The G-PEG-insulin conjugates were covalently bound to Con A that was attached to a PEG-poly(vinyl pyrrolidine-*co*-acrylic acid) backbone, and as the concentration of glucose increased competitive binding of glucose to Con A led to displacement and release of G-PEG insulin conjugates [135].

Other approach includes polymers with phenylboronic groups and polyol polymers that form a gel through complex formation between the pendant phenyl borate and hydroxyl groups [136]. Instead of polyol polymers, short molecules such as diglucosylhexadiamine have been used. As the glucose concentration increases, the cross-linking density of the gel decreases and as a result insulin is released from the eroded gel. The glucose exchange reaction is reversible and reformation of the gel occurs as a result of borate-polyol cross-linking. The major limitation of this system is the low specificity of PBA-containing polymers [137].

## KEYWORDS

- blood glucose levels
- carboxymethyl dextran
- concanavalin A
- dithiothreitol
- doxorubicin
- endogenous improved green fluorescence protein

## REFERENCES

1. You, J., Almeda, D., Ye, G., & Auguste, D., (2010). Bioresponsive matrices in drug delivery. *Journal of Biological Engineering, 4*(1), 15. doi: 10.1186/1751-1611-4-15.
2. Roy, I., & Gupta, M., (2003). Smart polymeric materials. *Chemistry and Biology, 10*(12), 1161–1171. doi: 10.1016/j.chembiol.2003.12.004.
3. Bawa, P., Pillay, V., Choonara, Y., & Du Toit, L., (2009). Stimuli-responsive polymers and their applications in drug delivery. *Biomedical Materials, 4*(2), 022001. doi: 10.1088/1741–6041/4/2/022001.
4. Priya, J. H., John, R., Alex, A., & Anoop, K., (2014). Smart polymers for the controlled delivery of drugs: A concise overview. *Acta Pharmaceutica Sinica B, 4*(2), 121–127. doi: 10.1016/j.apsb.2014.02.005.
5. Lu, Y., Aimetti, A., Langer, R., & Gu, Z., (2016). Bioresponsive materials. *Nature Reviews Materials, 2*(1), 16075. doi: 10.1038/natrevmats.2016.75.
6. Feng, S., Li, J., Luo, Y., et al., (2014). pH-sensitive nanomicelles for controlled and efficient drug delivery to human colorectal carcinoma lovo cells. *PLoS One, 9*(6), e100732. doi: 10.1371/journal.pone.0100732.
7. Liu, W., Yang, Y., Shen, P., et al., (2015). Facile and simple preparation of pH-sensitive chitosan-mesoporous silica nanoparticles for future breast cancer treatment. *Express Polymer Letters, 9*(12), 1061–1075. doi: 10.3144/expresspolymlett.2015.96.
8. Wu, G., Fang, Y., Yang, S., Lupton, J., & Turner, N., (2004). Glutathione metabolism and its implication for health. *The Journal of Nutrition, 134*, 481–492.
9. Kuppusamy, P., Li, H., Ilangovan, G., et al., (2002). Noninvasive imaging of tumor redox status and its modification by tissue glutathione levels. *Cancer Research, 62*, 301–312.
10. Meng, F., Hennink, W., & Zhong, Z., (2009). Reduction-sensitive polymers and bioconjugates for biomedical applications. *Biomaterials, 30*(12), 2181–2198. doi: 10.1016/j.biomaterials.2009.01.026.
11. Zhao, M., Biswas, A., Hu, B., et al., (2011). Redox-responsive nanocapsules for intracellular protein delivery. *Biomaterials, 32*(22), 5221–5230. doi: 10.1016/j.biomaterials.2011.03.060.

12. Miyata, K., Kakizawa, Y., Nishiyama, N., et al., (2004). Block catiomer polyplexes with regulated densities of charge and disulfide cross-linking directed to enhance gene expression. *Journal of the American Chemical Society, 126*(8), 2351–2361. doi: 10.1021/ja0379666.

13. Rotruck, J., Pope, A., Ganther, H., Swanson, A., Hafeman, D., & Hoekstra, W., (1973). Selenium: Biochemical role as a component of glutathione peroxidase. *Science, 179*(4073), 581–590. doi: 10.1126/science.179.4073.588.

14. Cao, W., Wang, L., & Xu, H., (2015). Selenium/tellurium containing polymer materials in nanobiotechnology. *Nano Today, 10*(6), 711–736. doi: 10.1016/j.nantod. 2015.11.004.

15. Ma, N., Li, Y., Xu, H., Wang, Z., & Zhang, X., (2010). Dual redox responsive assemblies formed from diselenide block copolymers. *Journal of the American Chemical Society, 132*(2), 441–443. doi: 10.1021/ja908124g.

16. Yang, J., Liu, W., Sui, M., Tang, J., & Shen, Y., (2011). Platinum (IV)-coordinate polymers as intracellular reduction-responsive backbone-type conjugates for cancer drug delivery. *Biomaterials, 32*(34), 9131–9143. doi: 10.1016/j.biomaterials.2011.08.022.

17. Levine, M., & Raines, R., (2012). Trimethyl lock: A trigger for molecular release in chemistry, biology, and pharmacology. *Chemical Science, 3*(8), 2412. doi: 10.1039/c2sc20536j.

18. Napoli, A., Valentini, M., Tirelli, N., Müller, M., & Hubbell, J., (2004). Oxidation-responsive polymeric vesicles. *Nature Materials, 3*(3), 181–189. doi: 10.1038/nmat1081.

19. Shim, M., & Xia, Y., (2013). A reactive oxygen species (ROS)-responsive polymer for safe, efficient, and targeted gene delivery in cancer cells. *Angewandte Chemie International Edition, 52*(27), 6921–6929. doi: 10.1002/anie.201209633.

20. Ma, Y., Dong, W., Hempenius, M., Möhwald, H., & Julius, V. G., (2006). Redox-controlled molecular permeability of composite-wall microcapsules. *Nature Materials, 5*(9), 721–729. doi: 10.1038/nmat1716.

21. Broaders, K., Grandhe, S., & Fréchet, J., (2011). A biocompatible oxidation-triggered carrier polymer with potential in therapeutics. *Journal of the American Chemical Society, 133*(4), 751–758. doi: 10.1021/ja110468v.

22. Noh, J., Kwon, B., Han, E., et al., (2015). Amplification of oxidative stress by a dual stimuli-responsive hybrid drug enhances cancer cell death. *Nature Communications, 6*(6907). doi: 10.1038/ncomms7907.

23. Liu, X., Xiang, J., Zhu, D., et al., (2016). Gene delivery: Fusogenic reactive oxygen species triggered charge-reversal vector for effective gene delivery (Adv. Mater. 9/2016). *Advanced Materials, 28*(9), 1711–1714. doi: 10.1002/adma.201670057.

24. Wang, M., Sun, S., Neufeld, C., Perez-Ramirez, B., & Xu, Q., (2014). Reactive oxygen species-responsive protein modification and its intracellular delivery for targeted cancer therapy. *Angewandte Chemie., 126*(49), 13661–13666. doi: 10.1002/ange.201407234.

25. Chung, M., Chia, W., Wan, W., Lin, Y., & Sung, H., (2015). Controlled release of an anti-inflammatory drug using an ultrasensitive ros-responsive gas-generating carrier for localized inflammation inhibition. *Journal of the American Chemical Society, 137*(39), 12461–12465. doi: 10.1021/jacs.5b08057.

26. Aran, K., Parades, J., Rafi, M., et al., (2015). Biosensors: Stimuli-responsive electrodes detect oxidative stress and liver injury (Adv. Mater. 8/2015). *Advanced Materials, 27*(8), 1431–1432. doi: 10.1002/adma.201570054.

27. Nsti, B. P., Vicente, M., Cabrera-García, A., & Fabregat, K., (2014). *Nanotechnology*. Place of publication not identified: CRC Press.

28. Cho, H., Bae, J., Garripelli, V., Anderson, J., Jun, H., & Jo, S., (2012). Redox-sensitive polymeric nanoparticles for drug delivery. *Chemical Communications, 48*(48), 6043. doi:10.1039/c2cc31463k.

29. Chang, B., Chen, D., Wang, Y., et al., (2013). Bioresponsive controlled drug release based on mesoporous silica nanoparticles coated with reductively sheddable polymer shell. *Chemistry of Materials, 25*(4), 571–585. doi: 10.1021/cm3037197.

30. Overall, C., & Kleifeld, O., (2006). Tumor microenvironment—opinion: Validating matrix metalloproteinases as drug targets and anti-targets for cancer therapy. *Nature Reviews Cancer, 6*(3), 221–239. doi: 10.1038/nrc1821.

31. Olson, E., Jiang, T., Aguilera, T., et al., (2010). Activatable cell penetrating peptides linked to nanoparticles as dual probes for *in vivo* fluorescence and MR imaging of proteases. *Proceedings of the National Academy of Sciences, 107*(9), 4311–4316. doi: 10.1073/pnas.0910283107.

32. Callmann, C., Barback, C., Thompson, M., Hall, D., Mattrey, R., & Gianneschi, N., (2015). Therapeutic enzyme-responsive nanoparticles for targeted delivery and accumulation in tumors. *Advanced Materials, 27*(31), 4611–4615. doi: 10.1002/adma.201501803.

33. Jiang, T., Olson, E., Nguyen, Q., Roy, M., Jennings, P., & Tsien, R., (2004). Tumor imaging by means of proteolytic activation of cell-penetrating peptides. *Proceedings of the National Academy of Sciences, 101*(51), 17861–17872. doi: 10.1073/pnas. 0408191101.

34. Nguyen, Q., Olson, E., Aguilera, T., et al., (2010). Surgery with molecular fluorescence imaging using activatable cell-penetrating peptides decreases residual cancer and improves survival. *Proceedings of the National Academy of Sciences, 107*(9), 4311–4322. doi: 10.1073/pnas.0910261107.

35. Zhang, S., Ermann, J., Succi, M., et al., (2015). An inflammation-targeting hydrogel for local drug delivery in inflammatory bowel disease. *Science Translational Medicine*, 7(300), 300ra121-300ra128. doi: 10.1126/scitranslmed.aaa5657.

36. Gajanayake, T., Olariu, R., Leclere, F., et al., (2014). A single localized dose of enzyme-responsive hydrogel improves long-term survival of a vascularized composite allograft. *Science Translational Medicine, 6*(249), 249ra111-249ra110. doi: 10.1126/scitranslmed.3008778.

37. Kim, H., Zhang, K., Moore, L., & Ho, D., (2014). Diamond nanogel-embedded contact lenses mediate lysozyme-dependent therapeutic release. *ACS Nano, 8*(3), 2991–3005. doi: 10.1021/nn5002968.

38. Jiang, T., Mo, R., Bellotti, A., Zhou, J., & Gu, Z., (2014). Drug delivery: Gel-liposome-mediated co-delivery of anticancer membrane-associated proteins and small-molecule drugs for enhanced therapeutic efficacy (Adv. Funct. Mater. 16/2014). *Advanced Functional Materials, 24*(16), 2251–2258. doi: 10.1002/adfm.201470099.

39. Hu, Q., Sun, W., Lu, Y., et al., (2016). Tumor microenvironment-mediated construction and deconstruction of extracellular drug-delivery depots. *Nano Letters, 16*(2), 1111–1126. doi: 10.1021/acs.nanolett.5b04343.

40. Kang, J., Asai, D., Kim, J., et al., (2008). Design of polymeric carriers for cancer-specific gene targeting: Utilization of abnormal protein kinase Cα activation in cancer cells. *Journal of the American Chemical Society, 130*(45), 14901–14907. doi: 10.1021/ja805364s.

41. Gu, Z., Yan, M., Hu, B., et al., (2009). Protein nanocapsule weaved with enzymatically degradable polymeric network. *Nano Letters*, *9*(12), 4531–4538. doi: 10.1021/nl902935b.

42. Linderoth, L., Peters, G., Madsen, R., & Andresen, T., (2009). Drug delivery by an enzyme-mediated cyclization of a lipid prodrug with unique bilayer-formation properties. *Angewandte Chemie, 121*(10), 1851–1858. doi: 10.1002/ange.200805241.

43. Maitz, M., Freudenberg, U., Tsurkan, M., Fischer, M., Beyrich, T., & Werner, C., (2013). Bio-responsive polymer hydrogels homeostatically regulate blood coagulation. *Nature Communications*, *4*. doi: 10.1038/ncomms3168.

44. Biswas, A., Joo, K., Liu, J., et al., (2011). Endoprotease-mediated intracellular protein delivery using nanocapsules. *ACS Nano*, *5*(2), 1381–1394. doi: 10.1021/nn1031005.

45. Jiang, T., Sun, W., Zhu, Q., Burns, N. A., Khan, S. A., Mo, R., & Gu, Z., (2015). Furin-mediated sequential delivery of anticancer cytokine and small-molecule drug shuttled by graphene. *Adv. Mater, 27,* 1021–1028.

46. Chen, Z., Li, Z., Lin, Y., Yin, M., Ren, J., & Qu, X., (2013). Bioresponsive hyaluronic acid-capped mesoporous silica nanoparticles for targeted drug delivery. *Chemistry - A European Journal*, *19*(5), 1771–1783.

47. Maitz, M., Freudenberg, U., Tsurkan, M., Fischer, M., Beyrich, T., & Werner, C., (2013). Bio-responsive polymer hydrogels homeostatically regulate blood coagulation. *Nature Communications*, *4*. doi: 10.1038/ncomms3168; 10.1002/chem.201202038.

48. Garripelli, V., Kim, J., Son, S., Kim, W., Repka, M., & Jo, S., (2011). Matrix metallo-proteinase-sensitive thermogelling polymer for bioresponsive local drug delivery. *Acta Biomaterialia*, *7*(5), 1981–1992. doi: 10.1016/j.actbio.2011.02.005.

49. Mo, R., Jiang, T., Di, J., Tai, W., & Gu, Z., (2014). Emerging micro- and nanotechnology based synthetic approaches for insulin delivery. *Chemical Society Reviews, 43*(10), 3595. doi: 10.1039/c3cs60436e.

50. Ying, J., Zion, T., & Zarur, A., (2013). *Stimuli-Responsive Systems for Controlled Drug Delivery.*

51. Pai, C., Bae, Y., Mack, E., Wilson, D., & Kim, S., (1992). Concanavalin a microspheres for a self-regulating insulin delivery system. *Journal of Pharmaceutical Sciences*, *81*(6), 531–536. doi: 10.1002/jps.2600810612.

52. Matsumoto, A., Ishii, T., Nishida, J., Matsumoto, H., Kataoka, K., & Miyahara, Y., (2011). Innentitelbild: A Synthetic approach toward a self-regulated insulin delivery system (Angew. Chem. 9/2012). *Angewandte Chemie., 124*(9), 2011–2018. doi: 10.1002/ange.201108617.

53. Makino, K., Mack, E., Okano, T., & Sung, W. K., (1990). A microcapsule self-regulating delivery system for insulin. *Journal of Controlled Release, 12*(3), 231–239. doi: 10.1016/0161-3659(90)90101-2.

54. Podual, K., Doyle, F., & Peppas, N., (2000). Glucose-sensitivity of glucose oxidase-containing cationic copolymer hydrogels having poly(ethylene glycol) grafts. *Journal of Controlled Release*, 67(1), 1–17. doi: 10.1016/s0161-3659(00)00191-4.

55. Chou, D., Webber, M., Tang, B., et al., (2015). Glucose-responsive insulin activity by covalent modification with aliphatic phenylboronic acid conjugates. *Proceedings of the National Academy of Sciences, 112*(8), 2401–2406. doi: 10.1073/pnas.1424684112.

56. Kim, H., Kang, Y., Kang, S., & Kim, K., (2012). Monosaccharide-responsive release of insulin from polymersomes of polyboroxole block copolymers at neutral pH. *Journal of the American Chemical Society, 134*(9), 4031–4033. doi: 10.1021/ja211728x.

57. Fischel-Ghodsian, F., Brown, L., Mathiowitz, E., Brandenburg, D., & Langer, R., (1988). Enzymatically controlled drug delivery. *Proc. Natl. Acad., 8,* 2401–2406.

58. Podual, K., (2000). Preparation and dynamic response of cationic copolymer hydrogels containing glucose oxidase. *Polymer, 41*(11), 3971–3983. doi: 10.1016/s0031-3861(99)00621-5.

59. Gu, Z., et al., (2013). Glucose-responsive microgels integrated with enzyme nanocapsules for closed-loop insulin delivery. *ACS Nano, 7,* 6758–6766.

60. Holtz, J. H., & Asher, S. A., (1997). Polymerized colloidal crystal hydrogel films as intelligent chemical sensing materials. *Nature, 389,* 829–832.

61. Goldraich, M., & Kost, J., (1993). Glucose-sensitive polymeric matrices for controlled drug delivery. *Clin. Mater. 13,* 135–142.

62. Gu, Z., et al., (2013). Injectable nano-network for glucose-mediated insulin delivery. *ACS Nano, 7,* 4194–4201.

63. Podual, K., Doyle, III, F. J., & Peppas, N. A., (2000). Dynamic behavior of glucose oxidase-containing microparticles of poly(ethylene glycol)-grafted cationic hydrogels in an environment of changing pH. *Biomaterials, 21,* 1439–1450.

64. Yu, J., Zhang, Y., Ye, Y., et al., (2015). Microneedle-array patches loaded with hypoxia-sensitive vesicles provides fast glucose-responsive insulin delivery. *Proceedings of the National Academy of Sciences, 112*(27), 8261–8265. doi: 10.1073/pnas.1505405112.

65. Yesilyurt, V., Webber, M., Appel, E., Godwin, C., Langer, R., & Anderson, D., (2015). Injectable self-healing glucose-responsive hydrogels with pH-regulated mechanical properties. *Advanced Materials, 28*(1), 81–91. doi: 10.1002/adma.201502902.

66. Liu, H., Li, Y., Sun, K., et al., (2013). Dual-responsive surfaces modified with phenylboronic acid-containing polymer brush to reversibly capture and release cancer cells. *Journal of the American Chemical Society, 135*(20), 7601–7609. doi: 10.1021/ja401000m.

67. Wang, C., Ye, Y., Hochu, G., Sadeghifar, H., & Gu, Z., (2016). Enhanced cancer immunotherapy by microneedle patch-assisted delivery of anti-PD1 antibody. *Nano Letters, 16*(4), 2331–2340. doi: 10.1021/acs.nanolett.5b05030.

68. Barone, P., & Strano, M., (2006). Reversible control of carbon nanotube aggregation for a glucose affinity sensor. *Angewandte Chemie, 118*(48), 8311–8321. doi: 10.1002/ange.200603138.

69. Yum, K., Ahn, J., McNicholas, T., et al., (2011). Boronic acid library for selective, reversible near-infrared fluorescence quenching of surfactant suspended single-walled carbon nanotubes in response to glucose. *ACS Nano, 6*(1), 811–830. doi: 10.1021/nn204323f.

70. Shibata, H., Heo, Y., Okitsu, T., Matsunaga, Y., Kawanishi, T., & Takeuchi, S., (2010). Injectable hydrogel microbeads for fluorescence-based *in vivo* continuous glucose monitoring. *Proceedings of the National Academy of Sciences, 107*(42), 17891–17898. doi: 10.1073/pnas.1006911107.

71. Veiseh, O., Tang, B., Whitehead, K., Anderson, D., & Langer, R., (2014). Managing diabetes with nanomedicine: Challenges and opportunities. *Nature Reviews Drug Discovery, 14*(1), 41–57. doi: 10.1038/nrd4477.

72. Sinha, A., Chakraborty, A., & Jana, N., (2014). Dextran-gated, multifunctional mesoporous nanoparticle for glucose-responsive and targeted drug delivery. *ACS Applied Materials and Interfaces, 6*(24), 22181–22191. doi: 10.1021/am505848p.

73. Zhao, L., Xiao, C., Wang, L., Gai, G., & Ding, J., (2016). Glucose-sensitive polymer nanoparticles for self-regulated drug delivery. *Chemical Communications, 52*(49), 7631–7652. doi: 10.1039/c6cc02202b.

74. Zhao, L., Huang, Q., Liu, Y., et al., (2017). Boronic acid as glucose-sensitive agent regulates drug delivery for diabetes treatment. *Materials, 10*(2), 170. doi: 10.3390/ma10020170.

75. Yoshida, T., Lai, T., Kwon, G., & Sako, K., (2013). pH- and ion-sensitive polymers for drug delivery. *Expert Opinion on Drug Delivery, 10*(11), 1491–1513. doi: 10.1517/17425247.2013.821978.

76. Seager, H., (1998). Drug-delivery products and the zydis fast-dissolving dosage form. *Journal of Pharmacy and Pharmacology, 50*(4), 371–382. doi: 10.1111/j.2041-7158.1998.tb06876.x.

77. Bodmeier, R., Guo, X., Sarabia, R., & Skultety, P., (1996). *Pharmaceutical Research, 13*(1), 51–56. doi: 10.1023/a:1016021115481.

78. Du, H., Wickramasinghe, R., & Qian, X., (2010). Effects of salt on the lower critical solution temperature of poly(N-Isopropylacrylamide). *The Journal of Physical Chemistry B, 114*(49), 16591–16604. doi: 10.1021/jp105652c.

79. Harada, A., & Kataoka, K., (1999). On-off control of enzymatic activity synchronizing with reversible formation of supramolecular assembly from enzyme and charged block copolymers. *Journal of the American Chemical Society, 121*(39), 9241–9242. doi: 10.1021/ja9919175.

80. Nakamura, T., Takashima, Y., Hashidzume, A., Yamaguchi, H., & Harada, A., (2014). A metal-ion-responsive adhesive material via switching of molecular recognition properties. *Nature Communications, 5*. doi: 10.1038/ncomms5622.

81. Yoshida, T., Shakushiro, K., & Sako, K., (2018). Ion-responsive drug delivery systems. *Current Drug Targets, 19*(3), 1–1. doi: 10.2174/1389450117666160527142138.

82. Kailash, C. G. (2015). Role of natural and functional polymers in drug delivery systems. *J. Pharma. Care Health Sys., 2*(5), 29.

83. Lao, Y., Phua, K., & Leong, K., (2015). Aptamer nanomedicine for cancer therapeutics: Barriers and potential for translation. *ACS Nano, 9*(3), 2231–2254. doi: 10.1021/nn507494p.

84. Naito, M., Ishii, T., Matsumoto, A., Miyata, K., Miyahara, Y., & Kataoka, K., (2012). A phenylboronate-functionalized polyion complex micelle for ATP-triggered release of siRNA. *Angewandte Chemie., 124*(43), 10901–10913. doi: 10.1002/ange.201203360.

85. Biswas, S., Kinbara, K., Niwa, T., et al., (2013). Biomolecular robotics for chemomechanically driven guest delivery fuelled by intracellular ATP. *Nature Chemistry, 5*(7), 611–620. doi: 10.1038/nchem.1681.

86. Mo, R., Jiang, T., DiSanto, R., Tai, W., & Gu, Z., (2014). ATP-triggered anticancer drug delivery. *Nature Communications*, 5. doi: 10.1038/ncomms4364.

87. Wu, C., Chen, T., Han, D., et al., (2013). Engineering of switchable aptamer micelle flares for molecular imaging in living cells. *ACS Nano, 7*(7), 5721–5731. doi: 10.1021/nn402517v.

88. Zhang, P., Wang, C., Zhao, J., et al., (2016). Near infrared-guided smart nanocarriers for microrna-controlled release of doxorubicin/SIRNA with intracellular ATP as fuel. *ACS Nano, 10*(3), 3631–3647. doi: 10.1021/acsnano.5b08145.

89. Sun, W., & Gu, Z., (2016). ATP-responsive drug delivery systems. *Expert Opinion on Drug Delivery, 13*(3), 311–314. doi: 10.1517/17425247.2016.1140147.

90. Harris, A., (2002). Hypoxia—a key regulatory factor in tumor growth. *Nature Reviews Cancer, 2*(1), 31–47. doi: 10.1038/nrc704.

91. Wilson, W., & Hay, M., (2011). Targeting hypoxia in cancer therapy. *Nature Reviews Cancer, 11*(6), 391–410. doi: 10.1038/nrc3064.

92. Perche, F., Biswas, S., Wang, T., Zhu, L., & Torchilin, V., (2014). Hypoxia-targeted siRNA delivery. *Angewandte Chemie., 126*(13), 3431–3434. doi: 10.1002/ange.201308368.

93. Zhang, G., Palmer, G., Dewhirst, M., & Fraser, C., (2009). A dual-emissive-materials design concept enables tumor hypoxia imaging. *Nature Materials, 8*(9), 741–751. doi: 10.1038/nmat2509.

94. Zheng, X., Wang, X., Mao, H., Wu, W., Liu, B., & Jiang, X., (2015). Hypoxia-specific ultrasensitive detection of tumors and cancer cells *in vivo*. *Nature Communications, 6*(5834). doi: 10.1038/ncomms6834.

95. Takasawa, M., Moustafa, R., & Baron, J., (2008). Applications of nitroimidazole *in vivo* hypoxia imaging in ischemic stroke. *Stroke, 39*(5), 1621–1637. doi: 10.1161/strokeaha.107.485938.

96. Kiyose, K., Hanaoka, K., Oushiki, D., et al., (2010). Hypoxia-sensitive fluorescent probes for *in vivo* real-time fluorescence imaging of acute ischemia. *Journal of the American Chemical Society, 132*(45), 15841–15848. doi: 10.1021/ja105937q.

97. Kulkarni, P., Haldar, M., You, S., Choi, Y., & Mallik, S., (2016). Hypoxia-responsive polymersomes for drug delivery to hypoxic pancreatic cancer cells. *Biomacromolecules, 17*(8), 2501–2513. doi: 10.1021/acs.biomac.6b00350.

98. Thambi, T., Deepagan, V., Yoon, H., et al., (2014). Hypoxia-responsive polymeric nanoparticles for tumor-targeted drug delivery. *Biomaterials, 35*(5), 1731–1743. doi: 10.1016/j.biomaterials.2013.11.022.

99. Roy, D., Brooks, W., & Sumerlin, B., (2013). New directions in thermoresponsive polymers. *Chemical Society Reviews, 42*(17), 7214. doi: 10.1039/c3cs35499g.

100. Yoshida, R., Uchida, K., Kaneko, Y., et al., (1995). Comb-type grafted hydrogels with rapid deswelling response to temperature changes. *Nature, 374*(6519), 241–242. doi: 10.1038/374240a0.

101. Huffman, A., Afrassiabi, A., & Dong, L., (1986). Thermally reversible hydrogels: II. Delivery and selective removal of substances from aqueous solutions. *Journal of Controlled Release, 4*(3), 211–222. doi: 10.1016/0161-3659(86)90001-2.

102. Bae, Y., Okano, T., Hsu, R., & Kim, S., (1987). *Die Makromolekulare Chemie, Rapid Communications, 8*(10), 481–485. doi: 10.1002/marc.1987.030081002.

103. Wang, C., Flynn, N., & Langer, R., (2004). Controlled structure and properties of thermoresponsive nanoparticle-hydrogel composites. *Advanced Materials, 16*(13), 1071–1079. doi: 10.1002/adma.200306516.

104. Oneal, D., (2004). Photo-thermal tumor ablation in mice using near infrared-absorbing nanoparticles. *Cancer Letters, 209*(2), 171–176. doi: 10.1016/s0301-3835(04)00141-2.

105. Timko, B., & Kohane, D., (2014). Prospects for near-infrared technology in remotely triggered drug delivery. *Expert Opinion on Drug Delivery, 11*(11), 1681–1685. doi: 10.1517/17425247.2014.930435.

106. Xia, L., Xie, R., Ju, X., Wang, W., Chen, Q., & Chu, L., (2013). Nano-structured smart hydrogels with rapid response and high elasticity. *Nature Communications, 4*. doi: 10.1038/ncomms3226.

107. Bae, Y., Okano, T., & Kim, S., (1989). Insulin permeation through thermo-sensitive hydrogels. *Journal of Controlled Release, 9*(3), 271–279. doi: 10.1016/0161-3659 (89)90091-5.

108. Okano, T., Bae, Y., Jacobs, H., & Kim, S., (1990). Thermally on-off switching polymers for drug permeation and release. *Journal of Controlled Release, 11*(1–3), 251–265. doi: 10.1016/0161-3659(90)90138-j.

109. Wang, C., Stewart, R., & Kopeček, J., (1999). Hybrid hydrogels assembled from synthetic polymers and coiled-coil protein domains. *Nature, 397*(6718), 411–420. doi: 10.1038/17092.

110. Di, J., Yao, S., Ye, Y., et al., (2015). Stretch-triggered drug delivery from wearable elastomer films containing therapeutic depots. *ACS Nano, 9*(9), 9401–9415. doi: 10.1021/acsnano.5b03975.

111. Ward, M. A., & Georgiou, T. K., (2011). *Thermoresponsive Polymers for Biomedical Applications of Polymers, 3*, 1215–1242.

112. Kikuchi, A., & Okano, T., (2002). Intelligent thermoresponsive polymeric stationary phases for aqueous chromatography of biological compounds. *Progress in Polymer Science, 27*, 1165–1193.

113. Hoffman, A. S., Stayton, P. S., Bulmus, V., Chen, G. H., Chen, J. P., & Cheung, C., (2000). Really smart bioconjugates of smart polymers and receptor proteins. *Journal of Biomedical Materials Research, 52*, 577–586.

114. Rein, V. U., Nurguse, B., Vineetha, J., Paul, D. T., Simon, J. T., Robert, J. M., Andrew, M. S., & Julie, E. G., (2007). *Bioresponsive Hydrogels, 10*(4), 41–48. https://doi.org/10.1016/S1369-7021(07)70049-4 (accessed on 16 January 2020).

115. Edinger, D., & Wagner, E., (2011). Bioresponsive polymers for the delivery of therapeutic nucleic acids: Wiley interdisciplinary reviews. *Nanomedicine and Nanobiotechnology, 3*(1), 31–46. doi: 10.1002/wnan.97.

116. Rahbek, U. L., Nielsen, A. F., Dong, M., You, Y., Chauchereau, A., Oupicky, D., Besenbacher, F., Kjems, J., & Howard, K. A., (2010). Bioresponsive hyperbranched polymers for siRNA and miRNA delivery. *Journal of Drug Target, 18*(10), 811–820. doi: 10.3109/1061186X.2010.527982.Epub2010Oct27.

117. Haijun, Y., & Ernst, W., (2009). Bioresponsive polymers for nonviral gene delivery. *Current Opinion in Molecular Therapeutics, 11*(2), 161–178.

118. Al-Tahami, K., & Singh, J., (2007). Smart polymer based delivery systems for peptides and proteins. *Recent Patents on Drug Delivery and Formulation, 1*, 65–71.

119. Kumar, A., Srivasthava, A., Galevey, I. Y., & Mattiasson, B., (2007). Smart polymers: Physical forms and bioengineering applications. *Progress in Polymer Science, 32*, 1205–1237.

120. Bawa, P., Viness, B., Yahya, E. C., & Lisa, C., (2009). Stimuli-responsive polymers and their applications in drug delivery. *Biomedical Materials, 4*, 022001.

121. Diez-Pena, E., Quijada-Garrido, I., & Barrales-Rienda, J. M., (2002). On the water swelling behaviour of poly(*N*-isopropylacrylamide) [P(*N*-iPAAm)], poly(methacrylic acid) [P(MAA)], their random copolymers and sequential interpenetrating polymer networks (IPNs). *Polymer, 43*, 4341–4348.

122. Varga, I., Gilanyi, T., Meszaros, R., Filipcsei, G., & Zrinyi, M., (2001). Effect of cross-link density on the internal structure of poly(*N*-isopropylacrylamide) microgels. *The Journal of Physical Chemistry B, 105*, 9071–9076.

123. Singh, S., Webster, D. C., & Singh, J., (2007). Thermosensitive polymers: Synthesis, characterization, and delivery of proteins. *International Journal of Pharmaceutics, 341*, 68–77.

124. Jeong, B., Kim, S. W., & Bae, W. H., (2002). Thermosensitive sol-gel reversible hydrogels. *Advanced Drug Delivery Review, 54*, 37–51.

125. Qiu, Y., & Park, K., (2001). Environment-sensitive hydrogels for drug delivery. *Advanced Drug Delivery Review, 53*, 321–339.

126. Aguilar, M. R., Elvira, C., Gallardo, A., Vázquez, B., & Román, J. S., (2007). Smart polymers and their applications as biomaterials. In: Ashammakhi, N., Reis, R., & Chiellini, E., (eds.), *Topics in Tissue Engineering* (Vol. 3, pp. 1–27).

127. Choi, S., Baudys, M., & Kim, S. W., (2004). Control of blood glucose by novel GLP-1 delivery using biodegradable triblock copolymer of PLGA-PEG-PLGA in type 2 diabetic rats. *Pharmaceutical Research, 21*, 827–831.

128. Ruel-Gariépy, E., & Leroux, J. C., (2004). *In situ*-forming hydrogels-review of temperature-sensitive systems. *European Journal of Pharmaceutics and Biopharmaceutics, 58*, 409–426.

129. Schmaljohann, D., (2006). Thermo-and pH-responsive polymers in drug delivery. *Advanced Drug Delivery Review, 58*, 1655–1670.

130. Chan, A., Orme, R. P., Fricker, R. A., & Roach, P., (2013). Remote and local control of stimuli responsive materials for therapeutic applications. *Advanced Drug Delivery Review, 65*, 497–514.

131. Foss, A. C., Goto, T., Morishita, M., & Peppas, N. A., (2000). Development of acrylic-based copolymers for oral insulin delivery. *European Journal of Pharmaceutics and Biopharmaceutics, 57*, 163–169.

132. Peng, C. L., Yang, L. Y., Luo, T. Y., Lai, P. S., Yang, S. J., & Lin, W. J., (2015). Development of pH sensitive 2-(disopropylamino) ethyl methacrylate based nanoparticles for photodynamic therapy. *Nanotechnology, 21*, 155013.

133. Gohy, J. F., Lohmeijer, B. G. G., Varshney, S. K., Decamps, B., Leroy, E., & Boileau, S., (2002). Stimuli-responsive aqueous micelles from an ABC metallo-supramolecular triblock copolymer. *Macromolecules, 35*, 9748–9755.

134. Ravaine, V., Ancla, C., & Catargi, B., (2008). Chemically controlled closed-loop insulin delivery. *Journal of Control Release, 132*, 2–11.

135. Roy, V. D., Cambre, J. N., & Sumerlin, B. S., (2010). Future perspectives and recent advances in stimuli-responsive materials. *Progress in Polymer Science, 35*, 278–301.

136. Takemoto, Y., Ajiro, H., Asoh, T. A., & Akashi, M., (2010). Fabrication of surface-modified. Hydrogels with polyion complex for controlled release. *Chemistry of Materials, 22*, 2923–2929.

137. Yin, R. X., Tong, Z., Yang, D. Z., & Nie, J., (2012). Glucose-responsive insulin delivery microhydrogels from methacrylated dextran/concanavalin A: Preparation and *in vitro* release study. *Carbohydrate Polymer, 89*, 117–123.

# CHAPTER 2

# Design of Bioresponsive Polymers

ANITA PATEL,[1] JAYVADAN K. PATEL,[1] and DEEPA H. PATEL[2]

[1]*Nootan Pharmacy College, Faculty of Pharmacy, Sankalchand Patel University, Visnagar–384315, Gujarat, India, E-mail: jayvadan04@gmail.com (J. K. Patel)*

[2]*Associate Professor, Department of Pharmaceutics, Parul Institute of Pharmacy and Research, Faculty of Pharmacy, Parul University, P.O. Limda, Ta. Waghodia, Vadodara–391760, Gujarat, India, Phone: 02668-260287, Fax: 02668-260201 E-mails: deepaben.patel@paruluniversity.ac.in, Pateldeepa18@yahoo.com*

## 2.1 SYNTHESIS OF BIO-RESPONSIVE POLYMERS

During the last few years, the technical and scientific significance of functional polymers has been well recognized; at present, a lot of concentration has been focused on bio-responsive polymers. Contrary to traditional polymers, to add in responsive components, it is essential to co-polymerize responsive blocks into a polymer or else copolymer backbone [1]. That's why, the formulation of distinct block copolymers with different architectures is vital: such as grafting of amphiphilic blocks into hydrophobic polymer backbone [2]. By means of living anionic polymerizations [3], cationic polymerizations [4] and controlled radical polymerizations (CRPs) methods [5], one can synthesize broad ranges of bio-responsive block copolymers. The increasing demand for definite and efficient soft resources in a range of nanoscale has directed to a momentous boost of events that merge architectural manage with the suppleness of integrating functional groups. Bearing in mind these considerations, there has been an important pursuit for clarifying various controlled polymerization approaches, which resulting into nitroxide-mediated radical polymerization (NMRP) [6, 7], atom transfer radical polymerization (ATRP) [8, 9], and reversible addition-fragmentation chain transfer (RAFT) processes [10, 11]. Ring-opening metathesis polymerization (ROMP) also presents a distinctive means of producing well-defined copolymers [12].

Additionally, to produce materials that can act in response to physiological signals, a budding drift in bio-engineering is to wangle substances that can interrelate with biological particulates. The synthesis of polymer and modification strategies is the present design approach. According to this approach, first polymers are synthesized by using different techniques and afterward synthesized polymer customized to make it responsive to biological response.

## 2.1.1  *LIVING RADICAL POLYMERIZATION (LRP)*

In the field of bio-responsive polymer synthesis, free radical polymerization has added a lot of attention attributable to its effortlessness, compatibility, and expediency [13]. As usual, the broad molecular weight distribution of the resultant polymers attributable to the termination process between two propagating radicals is the main restraint of conventional radical polymerization. Lately, ionic polymerization was the merely realistic route with respect to bio-responsive block copolymers with controlled molecular weight and design [14]. Given that ionic synthesis systems cannot be practical too numerous functional monomers and need meticulous keeping out of water and oxygen, LRP methods have been used to produce numerous stimuli-responsive copolymers with a variety of controlled designs [15]. The exploit of reagents which change chain propagating radicals into a "dormant" structure in equilibrium with the "active" form is a broad trait of these techniques.

ATRP, RAFT polymerization, and NMRP are the most ordinary techniques amid the LRP methods that have been urbanized [16]. The main characteristic of these techniques is the vibrant equilibrium among energetically propagating radicals as well as dormant polymer chains. Moreover, the states of reaction have to be chosen in such a way that the dormant moieties are preferential in the equilibrium which leads to importunate, low concentrations of transmitting radicals. A low concentration of propagating radical species is able to suppress the normal radical termination reactions successfully. Principally all technique is different in the chemistry of the limit resting on the dormant polymer chain. The caps used by ATPR, NMRP, and RAFT techniques are ω-halide; ω-alkoxamine and ω-dithioester in that order and ATRP give the impression to be the mainly imperative of the LRP techniques.

## 2.1.2   ATOM TRANSFER RADICAL POLYMERIZATION (ATRP)

ATRP technique is a controlled/living polymerization technique rooted in the utilization of radical polymerization for changing monomer into polymer. Matyjaszewski and coworkers were the first to use a simple, economical polymerization system for development of a controlled/living polymerization [17–19].

Through designing a proper catalyst, using an initiator with a suitable structure and fine-tuning the polymerization conditions ATRP was urbanized. In a view of this, during polymerization molecular weight augmented linearly with conversion and the polydispersities were distinctive for a living process [20]. The ATRP reaction mixture consisting of an initiator generally alkyl halogenides or chlorosulfonic acids, a transition metal catalyst, a ligand, a monomer, and if essential a solvent and other compounds such as an activator or a deactivator, as a result, is termed as a multi-component system. The equilibrium among dormant as well as active species is the important part determined by the selection of the appropriate catalyst/ligand system. Copper is used as the catalyst, although the use of ruthenium, rhodium, palladium, nickel, and iron have also been described in most studies [20]. Contingent on the metal center, the ligands are nitrogen or phosphor compounds with a wide structural diversity. The deactivator can be created *in situ*, otherwise for healthier manage; a minute quantity in relation to the catalyst can be included. Moreover, the catalyst is tolerant of small quantities of oxygen of as well as water and transition metals like iron and copper are employed as the catalysts. The ATRP method needs an extremely minute quantity of catalyst in contrast to other polymerization techniques.

The initiator should be selected in such a way that fast and quantitative initiation takes place. Allowing that, under these conditions, every polymer chains develop. For getting accurate control of the molecular weight and a low polydispersity index (PDI) this is one requirement. Distinctive alkyl halogenides as initiators acquire an acceptor substituent in the $\alpha$-position to the C–X bond to grow weaker the C–X bond. A quick and discriminating transfer of the halogen atom from the initiator to the metal center can be attained by doing this. In the majority cases, the halogen is chlorine or bromine.

The polymerization controls given by ATRP are an effect of the radical's formation which can grow; however, they are reversibly deactivated forming dormant species (see Figure 2.1) [20]. Dormant species reactivation permits on behalf of the polymer chains to grow up another time, merely deactivated afterward. That kind of progression consequences in a polymer chain that gradually, nevertheless progressively, grow, and have a distinct end group (generally, an alkyl halide is ending group for ATRP).

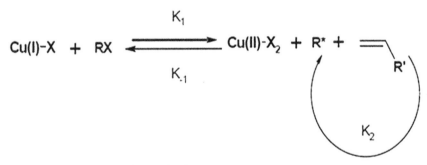

**FIGURE 2.1**   ATRP reaction mechanism.

The ATRP can be carried out in bulk as well as in a solution. If the polymer or catalyst complex is not soluble in the monomer, the use of a solvent is essential. Conversely, the solvent may have an effect on the ATRP process, and for the establishment of equilibrium between active and dormant species, altering the structure of the complex and improving its solubility, is fundamental. The rate and equilibrium of the transfer reaction can be determined by the structure of the complex. The transfer reaction betwixt the solvent and the growing radical should not occur to direct the polymerization. The foremost benefits of ATRP are the forbearance of various functional groups, allowing the polymerization of many monomers under proscribed conditions, for example, acrylamide, acrylate, acrylonitrile, methacrylate, styrene, and derivatives. At present, a range of attempts to have been ready to build up eco-friendly ATRP processes [21]. Through ATRP in water/2-propanol mixtures with a methoxy-polyethylene glycol, an amphiphilic random, gradient, and block copolymers of 2-dimethylaminoethyl methacrylate (DMAEMA) and *n*-butylmethacrylate were produced [22]. Poly(*N*-[(2,2-dimethyl-1,3-dioxolane)methyl]acrylamide), a new thermoresponsive polymer comprising pendant dioxolane groups, was synthesized by ATRP [23].

### 2.1.3 REVERSIBLE ADDITION-FRAGMENTATION CHAIN TRANSFER (RAFT)

The man-made couture of macro-molecules with multifaceted structural designs with a block, comb, graft, and star configurations with preset limited molecular weight distribution, molecular weight, and terminal functionality permitted by RAFT polymerization [24].

Subsequent to NMRP and ATRP, RAFT polymerization is the advanced living radical polymerization (LRP) method [25]. Through reversible addition and fragmentation chain transfer procedure, the free radical polymerization demonstrates living distinctiveness in the existence of RAFT agent. Homopolymers, plus block copolymers branched as well as ascent polymers with tapered polydispersities can effortlessly be organized via RAFT polymerization, [26]. In addition, it is entirely consistent with conventional free radical polymerization. The broad range of monomers that possibly will be simply polymerized with RAFT technique, primarily carboxyl monomers is the major benefit of this technique contrary to ATRP and NMRP polymerization [18, 19, 27].

Conventional radical polymerization entailed in the RAFT process in the existence of an apt chain transfer agent (CTA). Controlled chain growth is given by the degenerative transport between the growing radicals as well as the CTAs. An ample range of structurally diverse CTAs has been accounted for counting dithioesters, dithiocarbamates, dithiocarbonates (xanthates) and trithiocarbonates, [28]. The mechanism of the RAFT process is embodied in three major steps: initiation, propagation, and termination, the same as that of the conventional free-radical polymerization. The RAFT pre-equilibrium and the foremost RAFT equilibrium are two stages in RAFT propagation step [29]. The activation of the entire added CTA together with a little extent of propagation involved in the first stage, whereas chain equilibrium and propagation make up the second stage. The termination steps mostly concealed on account of the existence of the CTA and the succeeding degenerative transfer. The cautious choice of suitable monomers, initiators, and CTAs are the means to the structural organize in the RAFT process. While polymers are produced by means of ATRP, RAFT polymerization products are colored attributable to the thiocarbonylthio end groups. Conversely, this end groups can be easily detached via a post-treatment. An unsophisticated labeling system was detailed wherein the telechelic thiocarbonylthio functionality of distinct poly(*N*-isopropylacrylamides) (PNIPAAms) prepared by room-temperature RAFT polymerization was initial converted to the thiol and then reacted with a maleimido-functional fluorescent dye [30].

### 2.1.4   NITROXIDE MEDIATED RADICAL POLYMERIZATION (NMRP)

Stable free radicals, like nitroxides, are used as reversible terminating agents in the NMRP method, to manage the polymerization process [31, 32]. These radicals have a structure analogous to that of nitrogen monoxide. Through reversible deactivation of the growing chains, during covalent bond formation, dormant chains are produced. The single unpaired electron is delocalized over the nitrogen-oxygen bond, and this delocalization and the captodative structure of the radical ads to its permanence. The deactivation happens via the recombination of the radical chain ending by such stable nitroxide. The created C–O–N bond is thermolabile; moreover, the bond goes through a homolytic cleavage to make the active growing chain along with the nitroxide radical at high temperatures (90–130°C). The reaction temperature controls the equilibrium between active and dormant species. Activation is succeeded by a fast deactivation, where a handful of monomer units are integrated into the propagating chain. The development of a hydrido nitroxide is the latest foremost progress in nitroxide-mediated polymerization, wherein the existence of a hydrogen atom on the α-carbon results in a considerable rise in the range of vinyl monomers that go through controlled polymerization [33]. The beginning of the reaction can be attained via common initiators, for example, azobisisobutyronitrile or benzoyl peroxide. Iniferter is another approach in which the initiating as well as terminating moiety united in one molecule. By means of multifunctional iniferters, distinctive polymer structures such as block, star, or graft copolymers can be created [34, 35]. For instance, telechelic PNIPAAms could be produced through nitroxide-mediated controlled polymerization by introducing distinct end-group moieties. Through a triazole moiety different functional groups, connected to the central nitroxide-initiator arising from the so-called azide/alkyne-"click" reactions, were explored with isopropylacrylamides and butyl acrylate as monomers with regard to competence and livingness [36].

## 2.1.5 N-CARBOXY ANHYDRIDE (NCA) POLYMERIZATION

Long polypeptide chains are resourcefully synthesized by the polymerization of amino acid N-carboxyanhydrides [37, 38]. In Figure 2.2, the fundamental ring-opening mechanism of the NCA polymerization is demonstrated [38]. Simple reagents are used in NCA polymerization and it eases higher molecular weight polymers with high yield in uppermost purity and negligible racemization on the chiral centers [39]. The NCA polymerization method is an excellent option designed for a polypeptide having a high molecular weight (*HMW*) in contrast to the traditional solid-phase peptide synthesis methods in which bulky polypeptides (> 100 residues) grounding is not feasible because of the removal and reductions at some point in the deprotection and coupling steps [39].

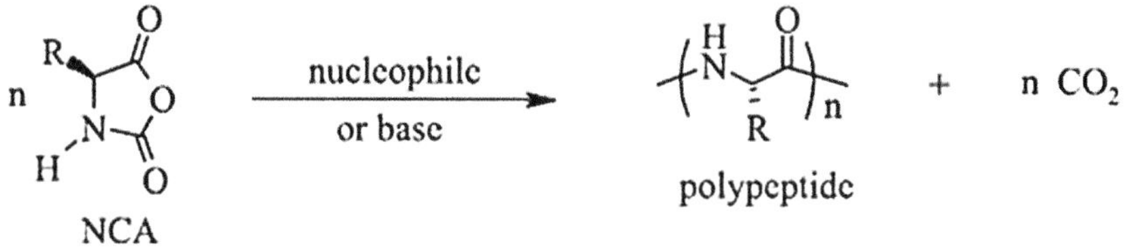

**FIGURE 2.2**    Simple NCA ring-opening polymerization mechanism.

## 2.1.6 RING-OPENING METATHESIS POLYMERIZATION (ROMP)

Industrially imperative products can be produced by the ROMP technique that is a kind of olefin metathesis chain-growth polymerization. The liberation of ring strain in cyclic olefins like cyclopentene or else norbornene is the instigator of the reaction and a broad variety of catalysts has been found.

In ROMP reaction, various metals used as the catalysts like simple ruthenium trichloride/alcohol mix to Grubbs' catalysts [40, 41]. The ROMP reaction is catalyzed mostly during the development of metal-carbene complexes when initially stated by Nobel Prize winner Yve Chauvin and Jean-Louis Hérisson [42, 43] even though a hydride system has as well described [40]. A beginning of the carbene species takes place in the course of various pathways; co-catalysts, substituent interactions, as well as solvent interactions, each, and everyone can give to the manufacture of the reactive catalytic species [40, 44, 45].

A strained cyclic structure is desired for the ROMP catalytic cycle as the source of power of the reaction is release of ring strain. Subsequent to development of the metal-carbene moieties, the carbene assaults the double bond in the constitution of ring shaping a greatly stressed metal-lacyclobutane intermediate. Subsequently, the ring unfolds providing the start of the polymer: by the use of a terminal double bond, a linear chain double bonded to the metal. The new-fangled carbene acts in response with the double bond on the subsequent monomer, accordingly promulgating the reaction [42].

Through the spatial functionalization of polymer constituents by specific biological entities, bio-responsive aptitude can be attained so as to stimulate the desired stimuli. Biological units used with this aspires can be native or synthetic bio-macromolecules, for instance, antibodies, enzymes, nucleic acids, [46] or small bioactive molecules like carbohydrates [47] or peptides.

Polymer networks can be shaped with couture their molecular structure that has bio-responsive properties. On the basis of following means, glucose-responsive polymeric based systems have been prepared: enzymatic oxidation of glucose via glucose oxidase (GOx), as well as conjugation of glucose with lectin or else reversible covalent bond formation with phenylboronic acid moieties. GOx is bound to a smart and pH-sensitive polymer for glucose-responsive polymers. Oxidation of glucose to gluconic acid is carried out by GOx, which brings about an alteration of pH in the environment [48]. Responding to the decreased pH, the pH-sensitive polymer after that shows a volume transition [49]. Oxidation-responsive vesicles as of amphiphilic block copolymers contingent on ethylene glycol plus propylene sulfide (PPS) rendering to oxidative circumstances were weakened [50]. Alteration of thioethers mostly in the hydrophobic PPS blocks was taken place within hydrophilic sulfoxides, have an effect on the hydrophilic-lipophilic balance of the amphiphile furthermore provoking its solubilization. The pH, as well as sugar-responsive behavior, has been shown by a polyethylene glycol-b-polystyrene boronic acid system with boronic acid (BA) moieties [51]. BA-containing polymers are prospective aspects for creating glucose-responsive materials given by the saccharide-sensitivity indued via the entrenched BA-diol interaction. Generally, phenylboronic acid with electron-withdrawing moieties is used [52, 53].

Disulfides convert into thiols in the existence of reducing agents, together with glutathione (GSH) in GSH-responsive systems, and the resultant thiol groups can reversibly restructuring disulfide bonds on oxidation. An attractive approach to build disulfide-containing materials rendered by a mild reaction condition of thiol-disulfide exchange [54]. Disulfides have as well been included in material systems in the form of disulfide-containing cross-linkers [55, 56].

Polymersomes on the basis of hydrophilic polyethylene glycol (PEG) and hydrophobic PPS linked via a disulfide bridge, cysteine (Cys) being there was disrupted PEG17-SS-PPS30, at a concentration in proportion to the intracellular level [57]. An analogous arrangement, on the basis of PEG-PPS block copolymers, was described previously [58]. The employment of oxidation was the first case in the existence of hydrogen peroxide ($H_2O_2$) to destabilize PEG-PPS-PEG vesicles moreover oxidizes the central-block sulfide species into sulphoxides and at last to sulphones, such oxidation bringing about an augment in the hydrophilicity of the firstly hydrophobic central block.

Reactive oxygen species like $H_2O_2$ and hydroxyl radicals are mostly targeted by oxidation-responsive materials, and a majority of oxidation-responsive resources are sulfur-based. Investigators copolymerized oxidation-convertible PPS by means of PEG to shape amphiphiles competent of self-assembling [58].

Enzyme-responsive polymeric systems can be formed through both using enzyme-degradable polymers as well as via modifying polymers with moieties which are receptive to precise enzymes. For enzyme-responsive systems, a majority of polymeric materials used comprise natural polymers, for example, chitosan (CS), pectin, dextran, and cyclodextrin, in addition to synthetic polymers, like PNIPAAm, polylysine (PL) and PEG.

To prepare enzyme-sensitive drug carriers α-chymotrypsin (hydrolase) as an enzyme has been used. For the harmonized delivery of deoxyribonucleic acid (DNA) and dextran, layer-by-layer self-assembly of DNA, as well as PL on porous calcium carbonate microparticles, has been produced. For the synchronized delivery of DNA and dextran, this dual-carrier has been used as a model system and with appropriate selection of both therapeutics; and also can be used for enhancing the therapeutic effectiveness of diseases [59]. The matrix metalloproteinases (MMPs) is one more imperative class of enzymes extensively used for a responsive system. The MMPs are a class of endopeptidases that exclusively cleave peptide bonds among non-terminal amino acids and are over-expressed in inflammation as well as cancer. Inclusion of peptide substrates appropriate for MMPs into the polymer can be used as enzyme cleavable sites furthermore assist in modifying the structure and morphology of the polymer system [60–62].

In recent times, disease-associated enzyme dysregulations have turned into a promising target because of the wide-ranging roles of enzymes in different biological processes. For example, ester bonds are often incorporated for targeting phosphatases, intracellular acid hydrolases and several other esterases; amides, although relatively stable to chemical attack in physiological environments, are vulnerable to enzymatic digestion and have been used for constructing materials sensitive to hydrolytic proteases, such as prostate-specific antigens; and materials containing cleavable azo linkers can target bacterial enzymes in the colon for site-specific drug release.

## 2.2   CHARACTERIZATION OF BIO-RESPONSIVE POLYMERS

Seeing as the length scale of bio-responsive polymers starting from unimers to micelles and finally, gels, some order of magnitudes a broad range of methods are desirable to describe them [63]. For the determination of the magnitude of polymerization, nuclear magnetic resonance (NMR) is used, whereas the molecular weight and polydispersity of the polymer can be found out by gel permeation chromatography (GPC). For visualization of the micellar structures of copolymers cryogenic transmission electron microscopy (Cryo-TEM) can be employed. Laser light scattering techniques might be useful to find out average molecular weights, the gyration radius of the polymer macromolecules; and to establish the size as well as the weight of the polymer micellar solutions as well. To study polymers in solutions form and in periodic gel form as well as solid networks form small-angle scattering (SAS) technique like X-ray as well as neutron can be employed with exploiting the inhomogeneities of the polymer's scattering length density. Techniques like dynamic mechanical study, rheological analysis, and differential scanning calorimetry (DSC) can be employed to know the mechanical, thermal, as well as rheological characteristics of smart bio-responsive copolymer gels, solids, plus micellar morphologies.

## 2.2.1  SPECTROSCOPIC TECHNIQUES

For the determination of the compositions of the synthesized polymer NMR, the technique is employed. Concerning the flexibility of the polymer core micelles, this technique offers details about the mobility of the polymer chains in their micelle form. The NMR technique is used to inspect the phase separation persuaded by alterations in both pH and temperature [64, 65].

Fourier transform infrared (FTIR) spectroscopy technique has been used for the self-association of aqueous surfactants, to probe a lipid bilayer transition, and for examining alterations in hydrogen bonding in polymers. Additionally, shifts in the infrared spectra give information about the confirmation as well as bond state of the functional groups included in the transition. The FTIR-Attenuated total reflection (ATR) spectra of PNIPAAm alter noticeably close to the coil-globule transition temperature in this method. Above lower critical solution temperature (LCST), the polymer will tend to aggregate and after that precipitate at the bottom of ATR crystal, this faithfully points out the incident of phase separation, and in fact, the polymer will moreover aggregate to a little degree over LCST by the infrared (IR) transmittance measurements. As well, the widespread 2D IR correlation spectroscopy, described by Noda [66, 67], may perhaps be successfully practical to the assessment of IR spectra of polymer solutions [68].

Ultraviolet (UV)-visible spectroscopy and fluorescence spectroscopy might be employed to investigate particles, which discriminatorily solvated into microdomains created with amphiphilic block copolymers, to examine self-assembly [24, 25].

## 2.2.2  SEPARATION TECHNIQUES

A quick system meant for the parting of oligomeric as well as polymeric moieties is GPC, also called size exclusion chromatography that is based upon the division of variations in molecular size in solution. For finding out the molecular weight distribution of synthetic polymers, GPC is an appropriate technique. With a view to find out the quantity of sample emerging, at the last part of the column a concentration detector is situated [69]. Furthermore, detectors possibly will be used to constantly find out the molecular weight of moieties eluting as of the column. An amount of solvent flow is as well observed to make available ways of distinguishing the molecular size of the eluting moieties. If required additional separation systems like high performance, liquid chromatography can be employed [69].

### 2.2.3   SCATTERING TECHNIQUES

#### 2.2.3.1   LASER LIGHT SCATTERING

For weight average molecular weights in addition to gyration radius of macromolecules laser light scattering method is a resourceful characterization method [37, 38] that can also be practical to find out the size as well as molecular weights of micelles within the solution. The Rayleigh ratio, the angular difference of the surplus absolute scattering intensity, is a result of the particle size as well as concentration along with the incident radiation's wavelength. In the light scattering experimentations, block copolymer micelles act like colloidal particles, at sizes 10–100 nm, whither intraparticle hindrance grows to be important, while the particles are earlier tiny in contrast to the wavelength of light, ($d < \lambda/20$). Additionally, the conventional light scattering technique is employed to learn the self-assembly constitutions of amphiphilic block copolymers attributable to its sensitiveness to the intermolecular interactions. In general, the relative scattered intensity against scattering vector correlation is robust through regression methods to extrapolate dimensions of the elements [70].

Quasi-elastic light scattering (QELS) or photon correlation spectroscopy builds upon the observable fact that the quick variations in the re-radiated light are linked with the rate of diffusion of the scattering elements [71]. The dependency of time for the light scattered as of a little part of the solution, above a time, range from tenths of a microsecond to milliseconds is determined in dynamic light scattering. After that, these variations in the intensity of the scattered light are connected to the diffusion rate of particles in and out of the area being considered (Brownian motion), along with the statistics that can be studied to straightforwardly provide the diffusion coefficients of the scattering particles. A distribution of diffusion coefficients is observed while manifold species are there. For direct measurement of the effectual, geometry independent, hydrodynamic radius of the particles the diffusion rate is used [72]. Size distributions of dispersed particles can be characterized by QELS or dynamic light scattering in dilute solutions in the dimensions of 4–2500 nm.

### 2.2.3.2 SMALL-ANGLE X-RAY AND NEUTRON SCATTERING

For examining structures of length scale with nanometer size range 1–300 nm SAS is a commanding method [73]. The significant attribute of this technique is it's prospective for investigating the internal arrangement of disordered systems; also normally, the appliance of this technique is an inimitable means to get straight structural data on systems with arbitrary structure of density inhomogeneities on this large scale. For study arrangements of particles into a solution like micelles or else colloids and periodic systems i.e., micelle networks or else lamella SAS is frequently used. The SAS information is the outcome of inhomogeneities in the scattering length density of a substance and there is no requirement of the crystallinity of the substance [74]. The scattering length density is a result of electron density in small-angle X-ray scattering.

In the sample, the x-rays interrelate with the electrons and so the technique is susceptible toward electron density dissimilarities [75]. There are both laboratory instruments anchored in added conventional basis and synchrotron-based instrumentations for these x-ray measurements. The neutrons are interacting with the nuclei in the sample in the small-angle neutron scattering (SANS) technique, and the interaction relies on the real isotope [76]. Hydrogen, as well as deuterium, has incredibly dissimilar scattering lengths and this can be developed on the contrary difference measurements wherein it is common that fraction of the particle in solution is deuterated. Through mixing deuterated and protonated solvents one can alter the scattering length density of the solvent. The system has requirement of a neutron source that is a nuclear reactor otherwise an accelerator base spallation source and for that reason, the experimentations are carried out at large scale services.

The progress as well as production of third-generation synchrotron radiation and higher flux spallation neutron sources has augmented the effectiveness of the method for structural examination of resources [74, 75]. To the extent that data analysis is fretful, numerous techniques have been urbanized for block copolymer micelle systems [76, 77]. Guinier and Fournet earliest composed the basic ideology of SAS in a characteristic monograph [75]. The techniques mostly comprise non-linear least square fit of analytical models to the SAS data to depict the form factor like intra-micellar scattering along with the structure factor i.e., inter-micellar scattering.

Careful treatment of the data is entailed for the execution of these models and the use of balancing methods, like Cryo-TEM and NMR to take out actually important data [78]. X-ray, as well as neutron scattering methods, are harmonizing to each other, albeit the fundamental hypothesis is similar [79]. They propose countless practical variations that can be browbeaten practically. There is an awfully small difference in electron scattering length density among organic polymers and water matrix, at the same time as the nuclear scattering length density difference can be diverse with correcting the $D_2O/H_2O$ proportion in SANS experiments.

## 2.2.4   *CRYOGENIC TRANSMISSION ELECTRON MICROSCOPY (CRYO-TEM)*

The Cryo-TEM has turned out to be an essential means to achieve high-resolution direct imagery of complex liquids, explicitly liquids with structure approximating nanometers to micrometers. A system that has been urbanized over the years facilitates us to confine the nanostructure in its inhabitant condition of preset concentration as well as temperature. An extensive range of arrangements of biological, low-, and high-molecular-weight solutes and synthetic has been deliberated previously through the method. The straight visualization of micellar structures created by various block copolymers is one of the extensively studied systems [80]. The samples can be ready as of fast vitrifying emaciated films of an aqueous copolymer solution to guard the self-assembled systems there at ambient temperature [81]. Diverse morphologies such as cylindrical, spherical, and bilayer otherwise vesical micelle arrangements have been envisaged with cryo-TEM. For examination of 'Y-junctions' of cylindrical micelles shaped as of diblock copolymers, Jain and Bates used cryo-TEM. The 'Y-junction' structure had been expected hypothetically, however, have not observed experimentally earlier [82].

By means of cryo-TEM, a number of block copolymer micelles have been experiential. The Cryo-TEM image analysis has been used by Zheng and coworkers to establish the corona and core dimensions of polyethylene oxide-b-polybutadiene micelles [83]. To average the dimensions from hordes of spherical micelles created from dilute solutions of Pluronic® F127 imaging analysis techniques established by Lam and co-workers [84]. Won and coworkers used cryo-TEM imaging to study polyethylene oxide (PEO)-based block copolymers [85]. The chain of diblock copolymers explored subsists in numerous micellar morphologies counting vesicle bilayers, cylinders, as well as spheres.

## 2.2.5 POLYMER PHASE PROPERTIES

For examination of the phase properties of copolymers DSC, simple potentiometric titrations, and tube inversion technique have been used, while to study the beginning of micelle and gel formation of amphiphilic copolymers calorimetric techniques have been used. The tube inversion test is a simple and most widely used method to get the gelation temperature. The majority of investigators employ alike criterion for the dissimilarities betwixt solution and gel. The gel contents have to stay position for some minutes upon inversion to be considering a gel.

The beginning of micelle formation into water is habitually escorted via an endothermic first-order transition, according to the dehydration energy of the hydrophobic block [73]. This is consistent with the fact that micellization is driven by entropic ads of separating the hydrophobic block parts to a core area. Prior investigations on copolymers like Pluronic® copolymer micelle self-assembly recognized an experimental correlation among a concentration of copolymer, the hydrophilic/hydrophobic composition and a critical micellization temperature (CMT) [86, 87]. At the same time, as the development of Pluronic® micelles consequences in a simply notable transition, the start of macroscopic gelation comes forwards to contain merely negligible transition energy. For this reason, an easy tube inversion method is used to learn the macroscopic gelation as a function of temperature. To get the pKa, isoelectric point values of the polymers in aqueous solutions potentiometric titrations will be used, in that way to envisage its ionization behavior by a change in the pH [88].

## 2.2.6   *MECHANICAL PROPERTIES*

Many techniques have been used to investigate the mechanical properties of physical hydrogels produced from block copolymers. Also, mechanical, and rheology studies recommend an outstanding ways of studying the sol-gel transition temperature of the thermoreversible hydrogels [89]. Hydrogel's viscoelastic properties are greatly reliant on chemical environment, concentration, and temperature and these viscoelastic properties of the gel determine the drug delivery applications of hydrogels *in vivo*. The hydrogel's shear modulus value is signifying the structure of the system. Analysis of the rheological properties of Pluronic® copolymers [90, 91], poly L-lactide-polyethylene glycol/poly D-lactide-polyethylene glycol block copolymers [92], Pluronic®-g-polyacrylic acid [93] and PEG-g-polylactic acid-co-glycolic acid [16, 92] have been carried out. Because of assorted reasons, rheological instrumentation is used to find out yield stress, to help out envisage shelf life and strain sweep to verify the critical strain (smallest energy required to disturb structure, where the greater the critical strain the superior the systems are detached). For the establishment of the CMT as well as the gelation point of Pluronic® solutions, Wang, and co-workers used rheology [90]. In a little temperature range, the gelation point was noticeable with a 103 fold rise in the shear modulus. The shear modulus did not considerably augment with temperature upon gelation. A pulse shear meter is used by Wang and Johnston to determine the shear modulus of solutions of hydrogels created by Pluronic® [94]. This technique is particularly valuable for hydrogels for the reason that it is very non-invasive, as a result reducing disturbances to the process of gel formation. The heat transfers all the way through the polymer solutions subject to the gelation process kinetics. This outcome has momentous insinuations for the impending injection of drug delivery devices based on Pluronic®. Upon injection, a slow-forming gel may well cause a 'burst effect,' seeing that a low viscosity gel network does not acquire the zero-order release properties so as to formulate Pluronic® hydrogels an eye-catching delivery device [95].

## 2.2.7   *SWELLING MEASUREMENTS*

To determine the equilibrium of swollen materials in distilled water, the dried grafted polymer subjected to swelling for a different time interval and then weighed them using analytical balance until a constant weight is achieved [96–98]. The extent of swelling of the films (%) finds out in accordance with the formula:

$$Swelling (\%) = [(Ws - Wd)/Wd]*100 \qquad (1)$$

In which, *Ws*, and *Wd* are weights of the swollen and initial catheter respectively.

### 2.2.8 pH CRITICAL POINT

The critical pH point is finding out in a similar fashion as of a plot of percentage of swelling in diverse pH solutions by a buffer of citric acid and sodium phosphate. The reactive mechanism fibs in the branches, side chain groups, in addition to the crosslink of a polymer's chemical structure in pH-sensitive smart materials. Absorption, as well as adsorption of water, can take place concurrently in the polymer networks which hold weak acid otherwise base groups. Ionization of the acid and base pendant groups have been taken place as the entry of water into the polymer network and this observable fact is controlled by means of the ionic composition, ionic strength and solution's pH [99–101]. One of the most accepted monomers is acrylic acid so as to have been attached to diverse polymeric matrices, moreover, its polymer or copolymers with pH-sensitive reaction contain the potential to go through a more chemical reaction to make novel functional assemblies [102, 103].

### 2.3 MECHANISM OF BIO-RESPONSIVE POLYMERS

The common system whereby a bio-responsive polymer produces a response in a hydrogel:

• A biological agent gets in touch through the hydrogel that is "sensed" via a biorecognition moiety in the hydrogel during binding or catalysis.
• Transduction takes place through upsetting the thermodynamic equilibrium of a bio-responsive hydrogel during chemical potential formed by the recognition reaction.

- A response crops up while the hydrogel structure re-establishes thermodynamic equilibrium on the basis of the new conditions.

To epitomize this procedure think about the case somewhere a substrate (i.e., bioactive agent) disperses into a bio-responsive hydrogel by way of an immobilized enzyme (biorecognition species). And this enzyme will initially make complex and commence to catalyze the substrate termed as a biological event. The substrate catalysis rate (transduction) will be determined by the concentration, or else chemical potential, of the substrate there in the hydrogel. A product will be endlessly produced in anticipation of the thermodynamic equilibrium is re-established (response).

Briefly, a bio-responsive polymer undergoes an abrupt change in its physical properties in response to a small biological stimulus. The polymer containing a bio-responsive system can either participate in a chemical reaction or result in cleavage of a chemical bond. A bio-responsive polymer undergoes transduction all the way through the networks of an extremely hydrophilic monomer including species like ethers, thiols or hydroxyls along with ionizable groups for example amines and carboxyls. The reaction betwixt a biorecognition species and a biological agent of bio-responsive polymer taken place and it generates or transfers ionically active moieties to the interstitial spaces of the hydrogel for the employment of ionic transduction. The equilibrium in the system altered by the ionic moieties and persuades the flow of chemical, electrical, as well as mechanical phenomenon ensuing in collapse or swelling. On the other hand, to collapse or swell a hydrogel ready from a bio-responsive polymer in response to a biological agent reversible crosslinks can be used.

Some more specific mechanism has been discussed in this chapter for various bio-responsive polymers like glucose-responsive, enzyme responsive, inflammation responsive, antigen responsive, DNA responsive, redox-responsive, and pH-responsive polymers.

### 2.3.1   GLUCOSE RESPONSIVE POLYMER

An aptitude of the glucose-responsive hydrogel systems has to present self-regulated insulin release in reaction to blood glucose level, in this manner concentration of insulin controlled within a normal range [104]. The most familiar characteristics of this hydrogel system make use of biocatalysts or else immobilized enzymes, distinctively GOx [105–107]. When an enzyme is covalently joined to a smart polymer, environmental alterations bring about major alterations in the conformation of polymer which drastically influences the activity of enzyme along with substrate contact to the enzyme molecule, which is a common mechanism behind this. These biocatalysts take action as a result of catalyzing an enzymatic reaction in their soluble state furthermore the products of this enzymatic reaction followed by activates the gel's phase transition. Investigations have revealed that gel beads of the enzyme-biopolymer complex produced by conjugating the biopolymer CS with carbodiimide conjugation [105]. GOx develops the pH sensitivity of the polymer used to immobilize the enzyme for the delivery of insulin (see Figure 2.3) [108]. For the formation of gluconic acid, GOx oxidizes glucose which brings about a pH alter in the surroundings [48]. Volume transition has been established by the pH-sensitive hydrogel responding to the lower pH, which happens as a result of the formation of gluconic acid. Consequently, the body's glucose concentrations control the swelling ratio of the hydrogel [49].

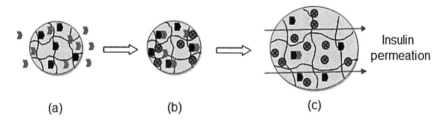

(a)  (b)  (c)

Glucose oxidase ( ■ ) Glucose ( ▮ ) Gluconic acid ( ⊗ )

**FIGURE 2.3** Schematic representation of a glucose-responsive glucose-oxidase-loaded membrane (a) glucose diffusion (b) enzymatic reaction and (c) swelling.

Polycations like poly(dimethylaminoethyl methacrylate) have been used for preparation of pH-sensitive membrane systems, in which a decline in pH resultant into swelling of the membrane which has a tendency to augment release of drug, e.g., insulin, this is commonly happening because of ionization of the polymer in the acidic medium [48, 109, 110]. With regards to the hydrogels are prepared using polyanions, the release of insulin controlled by diverse mechanisms. To shape grafted polyanion chains, these polyanions can be attached to a porous filter, like polymethacrylic acid-cobutyl methacrylate that is expanded at pH 7 and electrostatic repulsive forces amid the charges on the polymer chains are responsible for this. The chains breakdown when gluconic acid is formed, attributable to protonation of the carboxyl groups of the polymer which domino effect in the aperture of the pores along with diffusion of insulin [111]. To produce an insulin-loaded matrix, a dry pH-responsive polymer such as poly(dimethylaminoethyl methacrylate-*co*-ethyl acrylamide) was joined by bovine serum albumin (BSA), GOx, and insulin and finally compressed. Owing to the existence of GOx, the delivery system yielded gluconic acid by oxidation of glucose. This sequentially induced a decline in the pH, protonation, as well as swelling of the polymer, and finally insulin release.

A glucose-binding protein i.e., concanavalin A (Con A) is proficient in upholding four glucose units for each molecule. The competitive binding behavior of Con A with glucose and glycosylated insulin developed by the glycosylated insulin-Con A complex. The free glucose molecule brings about the displacement of glycosylated Con A-insulin couples within the surrounding tissues and is bioactive. Additional studies stated the synthesis of monosubstituted conjugates of glucosyl-terminal PEG as well as insulin. Covalent attachment took place between the G-PEG-insulin conjugates and Con A that was joined to a PEG-polyvinyl pyrrolidine-co-acrylic acid backbone, and when the concentration of glucose augmented competitive bonding of glucose with Con A directed to displacement and G-PEG insulin conjugates release [112, 113].

Swelling responding to higher glucose concentration was observed in hydrogels prepared from glucose-responsive polymer. The swelling of the hydrogels and therefore insulin release was driven by the complex produced flanked by glucose and phenylboronic acid [114].

### 2.3.2   ENZYME RESPONSIVE POLYMER

Naturally, bacteria mostly to be found in the colon generate particular enzymes, counting reductive enzymes like azoreductase or else hydrolytic enzymes like glycosidases which have the capability to degrade a variety of polysaccharides, for example, amylase/amylopectin, CS, cyclodextrin, dextrin, and pectin [115–117]. The polysaccharide dextran can be degraded by the microbial enzyme dextranase which developed by preparing hydrogels crosslinked with diisocyanate [118]. Enzymes are employed to demolish the polymer or its assemblies in the majority enzyme-responsive polymer systems and also these polymers do not necessitate an external stimulus for their breakdown, show high selectivity, and able to work in mild conditions. As an instance, polymer systems on the basis of alginate/CS or DEXS/CS microcapsules are reactive to chitosanase [119] and azoaromatic bonds are responsive to azoreductase [120]. To generate pH-responsive hydrogels azoaromatic bonds were employed as crosslinking agents owing to acidic co-monomers. As a result of the ionization of carboxylic acid groups, the hydrogels swell while passing through the GIT. Azoreductase can way in the cross-links of the swollen hydrogels and debase the matrix to release the protein drugs in the colon.

The enzyme-responsive polymers structure the base for hydrogels which are sensitive towards explicit enzymes. These enzymes are productively utilized as signals for the site-specific delivery of different drugs to specific organs and moreover are used as signals for observing quite a few physiological alterations. For targeting drugs particularly to the gastrointestinal tract the microbial inhabitants of the colon have been widely developed [121]. In a few words, enzymes induce hydrogel degradation and as a result the drug release by means of chemically controlled mechanisms [122].

### 2.3.3 INFLAMMATION-RESPONSIVE POLYMERS

T- and B-lymphocytes begins the inflammatory process, although amplified and perpetuated by means of macrophages and polymorphonuclear (PMN) leukocytes. In the process, a choice of chemical mediators, counting arachidonic acid metabolites, oxygen metabolites, and proteolytic enzymes can induce tissue damage. In the case of inflammation-responsive systems, the reactive oxygen metabolites, i.e., oxygen-free radicals free by macrophages and PMNs at some stage in the early phase of inflammation are the stimulus [123]. For a responsive drug delivery system, the aforementioned chemical mediators have been fruitfully employed as stimuli. As for instance, hyaluronic acid (HA) cross-linked with glycidyl ether can degrade responding to inflammation that revealed by the *in vivo* implantation experiments [124].

### 2.3.4 ANTIGEN RESPONSIVE POLYMER

The free antigen sensed by hydrogels formulated from antigen responsive polymer and undergoes swelling succeeded by drug release [108, 125].

### 2.3.5 DNA RESPONSIVE POLYMER

Swelling of single-stranded DNA grafted hydrogel probes exhibits in the company of single-stranded DNA [126].

### 2.3.6 REDOX RESPONSIVE POLYMER

In the reductive environment such as elevated GSH concentration in the intracellular matrix, disulfide linkages present in reduction sensitive hydrogels undergoes cleavage, and free bioactive molecules/drugs [127].

### 2.3.7 pH-RESPONSIVE POLYMERS

The presence of ionizable, weakly acidic or basic moieties, are the vital constituent for pH-responsive polymers that connect to a hydrophobic backbone, like polyelectrolytes [48, 128, 129]. The electrostatic repulsions of the produced anions or cations charges bring about a theatrical addition of coiled chains after ionization. As a result of the electrostatic effect as of other adjacent ionized groups, the ionization of the pendant acidic (e.g., carboxylic, and sulfonic acids) or basic (e.g., ammonium salts) groups on polyelectrolytes can be fractional [129, 130].

A different archetypal pH-responsive polymer demonstrates protonation/deprotonation episodes via giving out the charge above the ionizable groups of the molecule, like amino or carboxyl groups [131]. The degree of ionization, i.e., protonation or deprotonation, governs the swelling of pH-responsive hydrogels and Figure 2.4 explains that pH tempts a phase transition in pH-responsive polymers very unexpectedly. More often than not, the phase changes within 0.2–0.3 U of pH [132]. Pendant groups ionize and able to build up fixed charges on the polymer network in contact with aqueous media of suitable pH and ionic strength bringing about electrostatic repulsive forces accountable for pH-dependent swelling or else deswelling of the hydrogel, which at last regulates drug release [129].

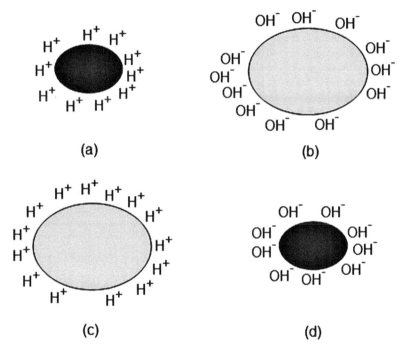

**FIGURE 2.4** The pH-responsive swelling of anionic and cationic hydrogels (a) non-swollen anionic hydrogel in an acidic environment, (b) swollen anionic hydrogel in the basic environment, (c) swollen cationic hydrogel in an acidic environment, and (d) non-swollen cationic hydrogel in a basic environment.

## KEYWORDS

- atom transfer radical polymerization
- attenuated total reflection
- chain transfer agent
- controlled radical polymerizations
- deoxyribonucleic acid
- Fourier transform infrared

## REFERENCES

1. Liu, R. X., Fraylich, M., & Saunders, B. R., (2009). *Thermoresponsive copolymers: From fundamental studies to applications. Colloid. Polym. Sci., 287,* 621–643.
2. Bosman, A. W., Vestberg, R., Heumann, A., Fréchet, J. M., & Hawker, C. J., (2003). A modular approach toward functionalized three-dimensional macromolecules: From synthetic concepts to practical applications. *J. Am. Chem. Soc., 125,* 711–728.
3. Rempp, P., & Lutz, P., (1993). Anionic polymerization methods: An efficient tool for the synthesis of tailor-made macromolecules. *Makromol. Chem. Macromoi. Symp., 67,* 1–14.
4. Kennedy, J. P., & Ivan, B., (1991). *Designed Polymers by Carbocationic Macromolecular Engineering, Theory, and Practice* (p. 9). Hanser Publishers, New York, and Munich.
5. Matyjaszewski, K., (1996). Controlled radical polymerization. *Curr. Opin. Solid State Mater Sci., 1,* 761–776.
6. Fischer, H., (1999). The persistent radical effect in controlled radical polymerizations. *J. Polym. Sci. A: Polym. Chem., 37,* 1881–1901.
7. Rodlert, M., Harth, E., Rees, I., & Hawker, C. J., (2000). End-group fidelity in nitroxide-mediated living free-radical polymerizations. *J. Polym. Sci. A: Polym. Chem., 38,* 4741–4763.
8. Matyjaszewski, K., (1995). Introduction to living polymerization. Living and/or controlled polymerization. *J. Phys. Org. Chem., 8,* 191–207.
9. Kato, M., Kamigaito, M., Sawamoto, M., & Higashimura, T., (1995). Polymerization of methyl methacrylate with the carbon tetrachloride/dichlorotris- (triphenylphosphine) ruthenium(II)/methylaluminum Bis(2,6-di-tert-butylphenoxide) initiating system: Possibility of living radical polymerization. *Macromol., 28,* 1721–1723.
10. Otsu, T., Matsumoto, A., & Tazaki, T., (1987). Radical polymerization of methyl methacrylate with some 1,2-disubstituted tetraphenylethanes as thermal iniferters. *Polym. Bull., 17,* 321–330.
11. Convertine, A. J., Ayres, N., Scales, C. W., Lowe, A. B., & McCornick, C. L., (2004). Facile, controlled, room-temperature RAFT polymerization of N-isopropylacrylamide. *Biomacromol., 5,* 1171–1180.
12. Trnka, T. M., & Grubbs, R. H., (2001). The development of L2X2Ru=CHR olefin metathesis catalysts: An organometallic success story. *Acc. Chem. Res., 43,* 11–29.
13. Wang, X. S., Jackson, R. A., & Armes, S. P., (2000). Facile synthesis of acidic copolymers via atom transfer radical polymerization in aqueous media at ambient temperature. *Macromol., 33,* 251–257.
14. Pyun, J., Kowalewski, T., & Matyjaszewski, K., (2003). Synthesis of polymer brushes using atom transfer radical polymerization. *Macromol. Rapid Commun., 24*(18), 1041–1059.
15. Wang, X. S., Lascelles, S. F., Jackson, R. A., & Armes, S. P., (1999). Facile synthesis of well-defined water-soluble polymers via atom transfer radical polymerization (ATRP) in aqueous media at ambient temperature. *Chem. Commun., 18,* 1811–1818.
16. Save, M., Weaver, J. V. M., Armes, S. P., & McKenna, P., (2002). Atom transfer radical polymerization of hydroxy-functional methacrylates at ambient temperature:

Comparison of glycerol monomethacrylate with 2-hydroxypropyl methacrylate. *Macromol., 35*(4), 1151–1159.

17. Link, A. J., & Tirrell, D. A., (2005). Reassignment of sense codons *in vivo. Methods, 36,* 291–298.

18. Wang, L., Brock, A., Herberich, B., & Schultz, P. G., (2001). Expanding the genetic code of *Escherichia coli. Science, 292,* 491–500.

19. Son, S., Tanrikulu, C., & Tirrell, D. A., (2006). Stabilization of bZIP peptides through incorporation of fluorinated aliphatic residues. *Chem. Bio. Chem., 7,* 1251–1257.

20. Matyjaszewski, K., & Xia, J., (2001). Atom transfer radical polymerization. *Chem. Rev., 101,* 2921–2990.

21. Tsarevsky, N. V., & Matyjaszewski, K., (2007). "Green" atom transfer radical polymerization: From process design to preparation of well-defined environmentally friendly polymeric materials. *Chem. Rev., 107,* 2271–2299.

22. Lee, S. B., Russell, A. J., & Matyjaszewski, K., (2003). ATRP synthesis of amphiphilic random, gradient, and block copolymers of 2-(dimethylamino)ethyl methacrylate and *n*-butyl methacrylate in aqueous media. *Biomacromol., 4,* 1381–1393.

23. Zou, Y., Brooks, D. E., & Kizhakkedathu, J. N., (2008). A novel functional polymer with tunable LCST. *Macromol., 41,* 5391–5405.

24. McCormick, C. L., Kirkland, S. E., & York, A. W., (2006). Synthetic routes to stimuli-responsive micelles, vesicles, and surfaces via controlled/living radical polymerization. *Polym. Rev., 46*(*4*), 421–443.

25. Wager, C. M., Haddleton, D. M., & Bon, S. A. F., (2004). A simple method to convert atom transfer radical polymerization (ATRP) initiators into reversible addition fragmentation chain-transfer (RAFT) mediators. *Eur. Polym. J., 40,* 641–645.

26. McCormick, C. L., & Lowe, A. B., (2004). Aqueous RAFT polymerization: Recent developments in synthesis of functional water-soluble (co)polymers with controlled structures. *Acc. Chem. Res., 37,* 311–325.

27. Holmgren, S. K., Taylor, K. M., Bretscher, L. E., & Raines, R. T., (1998). Code for collagen's stability deciphered. *Nature, 392,* 661–667.

28. Lowe, A. B., & McCormick, C. L., (2007). Reversible addition-fragmentation chain transfer (RAFT) radical polymerization and the synthesis of water-soluble (co) polymers under homogeneous conditions in organic and aqueous media. *Prog. Polym. Sci., 32,* 281–351.

29. Boyer, C., Bulmus, V., Davis, T. P., Ladmiral, V., Liu, J., & Perrier, S., (2009). Bioapplications of RAFT polymerization. *Chem. Rev., 109,* 5401–5436.

30. Scales, C. W., Convertine, A. J., & McCormick, C. L., (2006). Fluorescent labeling of RAFT-generated poly(N-isopropylacrylamide) via a facile maleimidethiol coupling reaction. *Biomacromol., 7*(5), 1381–1392.

31. Malmstrom, E. E., & Hawker, C. J., (1998). Macromolecular engineering via 'living' free radical polymerizations. *Macromol. Chem. Phys., 199,* 921–935.

32. Sciannamea, V., Jerome, R., & Detrembleur, C., (2008). *In-situ* nitroxide-mediated radical polymerization (NMP) processes: Their understanding and optimization. *Chem. Rev., 108*(3), 1101–1126.

33. Benoit, D., Hawker, C. J., Huang, E. E., Lin, Z., & Russell, T. P., (2000). One-step formation of functionalized block copolymers. *Macromol., 33*(5), 1501–1507.

34. Hawker, C. J., Bosman, A. W., & Harth, E., (2001). New polymer synthesis by nitroxide mediated living radical polymerizations. *Chem. Rev., 101,* 3661–3688.

35. Fukuda, T., Terauchi, T., Goto, A., Tsujii, Y., & Miyamoto, T., (1996). Well-defined block copolymers comprising styrene-acrylonitrile random copolymer sequences synthesized by "living" radical polymerization. *Macromol., 29,* 3051–3052.

36. Binder, W. H., Gloger, D., Weinstabl, H., Allmaier, G., & Pittenauer, E., (2007). Telechelic poly(*n*-isopropylacrylamides) via nitroxide-mediated controlled polymerization and *"click"* chemistry: Livingness and *"grafting-from"* methodology. *Macromol., 40,* 3091–3107.

37. Kricheldorf, H. R., (1987). *α-Aminoacid-N-Carboxyanhydrides and Related Materials.* Springer publishers, Berlin Heidelberg, New York.

38. Kricheldorf, H. R., (1990). In: Penczek, S., (eds.), *Models of Biolpolymers by Ring-Opening Polymerization.* CRC Press, Boca Raton, FL.

39. Deming, T. J., (2006). Polypeptide and polypeptide hybrid copolymer synthesis via NCA polymerization. *Adv. Polym. Sci., 202,* 1–18.

40. Mutch, A., Leconte, M., Lefebvre, F., & Basset, J. M., (1998). Effect of alcohols and epoxides on the rate of ROMP of norbornene by a ruthenium trichloride catalyst. *J. Mol. Catal. A: Chem., 133,* 191–199.

41. Scholl, M., Ding, S., Lee, C. W., & Grubbs, R. H., (1999). Synthesis and activity of a new generation of ruthenium-based olefin metathesis catalysts coordinated with 1,3-dimesityl-4,5-dihydroimidazol-2-ylidene ligands. *Org. Lett., 1,* 951–956.

42. Grubbs, R. H., & Tumas, W., (1989). Polymer synthesis and organotransition metal chemistry. *Science, 243,* 901–915.

43. Hérisson, J. L., & Chauvin, Y., (1971). Catalysis of transformation of olefins by tungsten complexes. II. Telomerization of cyclic olefins in the presence of acyclic olefins. *Die Makromolekul Chem., 141,* 161–176.

44. Slugovc, C., Demel, S., Riegler, S., Hobisch, J., & Stelzer, F., (2004). Influence of functional groups on ring opening metathesis polymerization and polymer properties. *J. Mol. Catal. A: Chem., 213,* 101–113.

45. Zhang, D., Huang, J., Qian, Y., & Chan, A. S. C., (1998). Ring-opening metathesis polymerization of norbornene and dicylopentadiene catalyzed by $C_{p2}TiC_{l2}/RMgX$. *J. Mol. Catal. A: Chem., 133,* 131–133.

46. Carlini, A. S., Adamiak, L., & Gianneschi, N. C., (2016). Biosynthetic polymers as functional materials. *Macromol., 49,* 4371–4394.

47. Russo, L., & Cipolla, L., (2016). Glycomics: New challenges and opportunities in regenerative medicine. *Chem. Eur. J., 22,* 13381–13388.

48. Qiu, Y., & Park, K., (2001). Environment-sensitive hydrogels for drug delivery. *Adv. Drug Deliv. Rev., 53, 3,* 321–339.

49. Chaterji, S., Kwon, I. K., & Park, K., (2007). Smart polymeric gels: Redefining the limits of biomedical devices. *Prog. Polym. Sci., 32*(8), 1081–1122.

50. Napoli, A., Boerakker, M. J., Tirelli, N., Nolte, R. J. M., Sommerdijk, N. A. J. M., & Hubbell, J. A., (2004). Glucose-oxidase based self-destructing polymeric vesicles. *Langmuir, 20,* 3481–3491.

51. Kim, K. T., Cornelissen, J. J. L. M., Nolte, R. J. M., & Van Hest, J. C. M., (2009). A polymersome nanoreactor with controllable permeability induced by stimuliresponsive blockcopolymers. *Adv. Mater, 21,* 2781–2791.

52. Matsumoto, A., Ishii, T., Nishida, J., Matsumoto, H., Kataoka, K., & Miyahara, Y., (2012). A synthetic approach toward a self-regulated insulin delivery system. *Angew. Chem. Int. Ed., 51,* 2121–2128.

53. Chou, D. H., Webber, M. J., Tang, B. C., Lin, A. B., Thapa, L. S., Deng, D., Truong, J. V., Cortinas, A. B., Langer, R., & Anderson, D. G., (2015). Glucose-responsive insulin activity by covalent modification with aliphatic phenylboronic acid conjugates. *Proc. Natl. Acad. Sci. USA, 112,* 2401–2406.

54. Meng, F., Hennink, W. E., & Zhong, Z., (2009). Reduction-sensitive polymers and bioconjugates for biomedical applications. *Biomaterials, 30,* 2181–2198.

55. Zhao, M., Biswas, A., Hu, B., Joo, K. I., Wang, P., Gu, Z., & Tang, Y., (2011). Redox-responsive nanocapsules for intracellular protein delivery. *Biomaterials, 32,* 5221–5230.

56. Miyata, K., Kakizawa, Y., Nishiyama, N., Harada, A., Yamasaki, Y., Koyama, H., & Kataoka, K., (2004). Block catiomer polyplexes with regulated densities of charge and disulfide cross-linking directed to enhance gene expression. *J. Am. Chem. Soc., 126,* 2351–2361.

57. Cerritelli, S., Velluto, D., & Hubbell, J. A., (2007). PEG-SS-PPS: Reduction-sensitive disulfides block copolymer vesicles for intracellular drug delivery. *Biomacromol., 8,* 1961–1972.

58. Napoli, A., Valentini, M., Tirelli, N., Muller, M., & Hubbell, J. A., (2004). Oxidation-responsive polymeric vesicles. *Nat. Mater, 3,* 181–189.

59. Wang, Z., Qian, L., Wang, X., Zhu, H., Yang, F., & Yang, X., (2009). Hollow DNA/ PLL microcapsules with tunable degradation property as efficient dual drug delivery vehicles by α-chymotrypsin degradation. *Colloids Surf. A., 332,* 161–171.

60. Chien, M. P., Thompson, M. P., Lin, E. C., & Gianneschi, N. C., (2012). Fluorogenic enzyme-responsive micellar nanoparticles. *Chem. Sci., 3,* 2691–2694.

61. Ku, T. H., Chien, M. P., Thompson, M. P., Sinkovits, R. S., Olson, N. H., Baker, T. S., & Gianneschi, N. C., (2011). Controlling and switching the morphology of micellar nanoparticles with enzymes. *J. Am. Chem. Soc., 133,* 8391–8395.

62. De Graaf, A. J., Mastrobattista, E., Vermonden, T., Van Nostrum, C. F., Rijkers, D. T., Liskamp, R. M., & Hennink, W. E., (2012). Thermosensitive peptide-hybrid ABC block copolymers obtained by ATRP: Synthesis, self-assembly, and enzymatic degradation. *Macromol., 45,* 841–851.

63. Alexandridis, P., & Lindman, B., (2000). *Amphiphilic Block Copolymers: Self-Assembly and Applications* (p. 305) Elsevier, New York.

64. Kirshenbaum, K. S., Isaac, C., & David, A. T., (2002). Biosynthesis of proteins incorporating a versatile set of phenylalanine analogues. *Chem. Biochem., 3,* 231–237.

65. Nowak, A. P., Breedveld, V., Pakstis, L., Ozbas, B., Pine, D. J., Pochan, D., & Deming, T. J., (2002). Rapidly recovering hydrogel scaffolds from self-assembling diblock copolypeptide amphiphiles. *Nature, 417,* 421–428.

66. Noda, I., Dowrey, A. E., Marcott, C., Story, G. M., & Ozaki, Y., (2000). Generalized two-dimensional correlation spectroscopy. *J. Appl. Spectrosc., 54,* 231–248.

67. Noda, I., (1993). Generalized two-dimensional correlation method applicable to infrared, Raman, and other types of spectroscopy. *J. Appl. Spectrosc., 47,* 1321–1336.

68. Guo, Y., Peng, Y., & Wu, P., (2008). A two-dimensional correlation ATR-FTIR study of poly(vinyl methyl ether) water solution. *J. Mol. Struct., 875,* 481–492.

69. Stanely, R. S., Wolf, K., Jo-Anne, B., & Eli, M. P., (1998). *Polymer Synthesis and Characterization Methods.* A Laboratory Manual.

70. Huglin, M. B., (1972). *Light Scattering From Polymer Solutions.* Academic Press, New York.

71. Charles, S. J. J., & Don, A. G., (1981). *Laser Light Scattering.* Dove Publications.

72. Berne, B., & Pecora, R., (2000). *Dynamic Light Scattering: With Applications to Chemistry, Biology and Physics.* Wiley-Interscience, New York.

73. Feigin, L. A., & Svergun, D. I., (1987). *Structure Analysis by Small-Angle X-Ray and Neutron Scattering.* Springer US, New York.

74. Roe, R. J., (2000). *Methods of X-Ray and Neutron Scattering in Polymer Science.* Oxford University Press, New York.

75. Guinier, A., & Fournet, G., (1985). *Small-Angle Scattering of X-Rays.* Wiley, New York.

76. Castelletto, V. A. I. W. H., (2003). Small-angle scattering functions of micelles. *Fiber Diffraction Review, 11,* 31–43.

77. McConnell, G. A., Gast, A. P., Huang, J. S., & Smith, S. D., (1993). Disorder order transitions in soft sphere polymer micelles. *Phys. Rev. Lett., 71,* 2101–2105.

78. Yu, K., & Eisenberg, A., (1998). Bilayer morphologies of self-assembled crew-cut aggregates of amphiphilic PS-b-PEO diblock copolymers in solution. *Macromol., 31,* 3501–3518.

79. Weyerich, B., Brunner-Popela, J., & Glarter, O., (1999). Bilayer morphologies of self-assembled crew-cut aggregates of amphiphilic PS-b-PEO diblock copolymers in solution. *J. Appl. Crystallogr., 32,* 191–209.

80. Gelhaaf, S. U., Schurtneberger, P., & Muller, M., (2000). New controlled environment vitrification system for cryo-trasmission electron microscopy: Design and application to surfactant solution. *J. Microsc., 200,* 121–139.

81. Wang, H., Wang, H. H., Urban, V. S., Littrell, K. C., Thiyagarajan, P., & Yu, L., (2000). Syntheses of amphiphilic diblock copolymers containing a conjugated block and their self-assembling properties. *J. Am. Chem. Soc., 122,* 6851–6861.

82. Jain, S., & Bates, F. S., (2003). On the origins of morphological complexity in block copolymer surfactants. *Science, 300,* 461–464.

83. Zheng, Y., Won, Y. Y., Bates, F. S., Davis, H. T., & Scriven, L. E., (1999). Directly resolved core-corona structure of block copolymer micelles by cryo-transmission electron microscopy. *J. Phys. Chem. B, 103,* 10331–10334.

84. Lam, Y. M., Grigorieff, N., & Goldbeck-Wood, G., (1999). Direct visualization of micelles of Pluronic block copolymers in aqueous solution by cryo-TEM. *Phys. Chem. Chem. Phys, 1,* 3331–3334.

85. Won, Y. Y., Brannan, A. K., Davis, H. T., & Bates, F. S., (2002). Cryogenic transmission electron microscopy (Cryo-TEM) of micelles and vesicles formed in water by poly(ethylene oxide)-based block copolymers. *J. Phys. Chem. B, 106,* 3351–3364.

86. Wanka, G., Hoffmann, H., & Ulbricht, W., (1994). Phase diagrams and aggregation behavior of poly(oxyethylene)-poly(oxypropylene)-poly(oxyethylene) triblock copolymers in aqueous solutions. *Macromol., 27,* 4141–4159.

87. Vadnere, M., Amidon, G., Lindenbaum, S. D., & Haslam, J. L., (1984). Thermodynamics studies on the gel-sol-transition of some Pluronic polyols. *Int. J. Pharm., 22,* 201–218.

88. Petrov, A. I., Antipov, A. A., & Sukhorukov, G. B., (2003). Base-acid equilibria in polyelectrolyte systems: From weak polyelectrolytes to interpolyelectrolyte complexes and multilayered polyelectrolyte shells. *Macromol., 36,* 10071–10086.

89. Pozzo, D. C., Hollabaugh, K. R., & Walker, L. M., (2005). Rheology and phase behavior of copolymer-templated nanocomposite materials. *J. Rheol., 49,* 751–782.

90. Wang, P. A. T. P. J., (1991). Kinetics of sol-to gel transition for polosamer polyols. *J. Appl. Polym. Sci., 43,* 281–292.

91. Nystroem, B., & Walderhaug, H., (1996). Dynamic viscoelasticity of an aqueous system of a poly(ethylene oxide)-poly(propylene oxide)-poly(ethylene oxide) triblock copolymer during gelation. *J. Phys. Chem., 100,* 5431–5439.

92. Fujiwara, T., Mukose, T., Yamaoka, T., Yamane, H., Sakurai, S., & Kimura, Y., (2001). Novel thermo-responsive formation of a hydrogel by stereocomplexation between PLLA-PEG-PLLA and PDLA-PEG-PDLA block copolymers. *Macromol. Biosci., 1,* 201–208.

93. Orkisz, M. J., (1997). *Rheological Properties of Reverse Thermogelling Poly(Acrylic Acid)-g-(Oxyethylene-b-Oxypropylene-b-Oxyethylene) Polymers (Smart Hydrogel).* Book of Abstracts, 213th ACS National Meeting, San Francisco.

94. Wang, P., & La, T. P. J., (1995). Sustained-release Interleukin-2 following intramuscular injection in rats. *Int. J. Pharm., 113,* 71–81.

95. Jeong, B., Wang, L. Q., & Gutowska, A., (2001). Biodegradable thermoreversible gelling PLGA-g-PEG copolymers. *Chem. Commun., 16,* 1511–1517.

96. Haraguchi, K., & Takehisa, T., (2002). Nanocomposite hydrogels: A unique organic-inorganic network structure with extraordinary mechanical, optical, and swelling/de-swelling properties. *Adv. Mater., 14,* 1121–1124.

97. Zhang, W. A., Luo, W., & Fang, Y. E., (2005). Synthesis and properties of a novel hydrogel nanocomposite. *Mater. Lett., 59,* 2871–2880.

98. Burillo, G., Bucio, E., Arenas, E., & Lopez, G. P., (2007). Temperature and pH sensitive swelling behavior of binary DMAEMA/4VP grafts on polypropylene films. *Macromol. Mater. Eng., 292,* 211–219.

99. Lowman, A. M., Morishita, M., Kajita, M., Nagai, T., & Peppas, N. A., (1999). Oral delivery of insulin using pH-responsive complexation gels. *J. Pharm. Sci., 88,* 931–937.

100. Wood, K. M., Stone, G., & Peppas, N. A., (2006). Lectin functionalized complexation hydrogels for oral protein delivery. *J. Control. Release, 116,* e66–e68.

101. Wood, K. M., Stone, G. M., & Peppas, N. A., (2008). Wheat germ agglutinin functionalized complexation hydrogels for oral insulin delivery. *Biomacromol., 9,* 1291–1298.

102. Bucio, E., & Burillo, G., (2007). Radiation grafting of pH and thermosensitive N-isopropylacrylamide and acrylic acid onto PTFE films by two-steps process. *Radiat. Phys. Chem., 76,* 1721–1727.

103. Huang, M., Jin, X., Li, Y., & Fang, Y., (2006). Muddied guar gum matrix tablet for controlled release of diltiazem hydrochloride. *React. Funct. Polym., 66,* 1041–1046.

104. Gil, E. S., & Hudson, S. M., (2004). Stimuli-responsive polymers and their bioconjugates *Prog. Polym. Sci., 29,* 1171–1222.

105. Ghanem, A., & Ghaly, A., (2004). Immobilization of glucose oxidase in chitosan gel beads *J. Appl. Polym. Sci., 91,* 861–866.

106. Srivastava, R., Brown, J. Q., Zhu, H., & McShane, M. J., (2005). Stabilization of glucose oxidase in alginate microspheres with photoreactive diazoresin nanofilm coatings *Biotechnol. Bioeng., 91,* 121–131.

107. Kang, S. I., & Bae, Y. H., (2003). A sulfonamide based glucose-responsive hydrogel with covalently immobilized glucose oxidase and catalase. *J. Control. Release, 86,* 111–121.

108. Miyata, T., Uragamia, T., & Nakamaeb, K., (2002). Biomolecule sensitive hydrogels. *Adv. Drug Deliv. Rev., 54,* 71–98.

109. Albin, G., Horbett, T. A., & Ratner, B. D., (1985). Glucose sensitive membranes for controlled delivery of insulin: Insulin transport studies. *J. Control. Release, 2,* 151–164.

110. Ishihara, K., Kobayashi, M., & Shinohara, I., (1984). Glucose induced permeation control of insulin through a complex membrane consisting of immobilized glucose oxidase and a poly(amine) *Polym. J., 16,* 621–631.

111. Ito, Y., Casolaro, M., Kono, K., & Yukio, I., (1989). An insulin-releasing system that is responsive to glucose. *J. Control. Release, 10,* 191–203.

112. Takemoto, Y., Ajiro, H., Asoh, T. A., & Akashi, M., (2010). Fabrication of surface-modified hydrogels with polyion complex for controlled release. *Chem. Mater, 22,* 2921–2929.

113. Kim, S. W., Pai, C. M., Makino, K., Seminoff, L. A., Holmberg, D. L., Gleeson, J. M., Wilson, D. E., & Mack, J. E., (1990). Self-regulated glycosylated insulin delivery. *J. Control. Release, 11,* 191–201.

114. Guenther, M., Wallmersperger, T., Keller, K., & Gerlach, G., (2013). Swelling behavior of functionalized hydrogels for application in chemical sensors. In: Sadowski, G., & Richtering, W., (eds.), *Intelligent Hydrogels* (pp. 261–273). Springer: Cham, Switzerland.

115. Chambin, O., Dupuis, G., Champion, D., Voilley, A., & Pourcelot, Y., (2006). Colon-specific drug delivery: Influence of solution reticulation properties upon pectin beads performance. *Int. J. Pharm., 321,* 81–93.

116. Sinha, V. R., & Kumria, R., (2001). Polysaccharides in colon-specific drug delivery. *Int. J. Pharm., 224,* 11–38.

117. Vandamme, T. F., Lenourry, A., Charrueau, C., & Chaumeil, J. C., (2002). The use of polysaccharides to target drugs to the colon. *Carbohydr. Polym., 48,* 211–231.

118. Hovgaard, L., & Brondsted, H., (1995). Dextran hydrogels for colon-specific drug delivery. *J. Control. Release, 36,* 151–166.

119. Itoh, Y., Matsusaki, M., Kida, T., & Akashi, M., (2006). Enzyme-responsive release of encapsulated proteins from biodegradable hollow capsules. *Biomacromol., 7,* 2711–2718.

120. Akala, E. O., Kopeckova, P., & Kopecek, J., (1998). Novel pH-sensitive hydrogels with adjustable swelling kinetics. *Biomaterials, 19,* 1031–1047.

121. Yang, L., Chu, J., & Fix, J., (2002). Colon-specific drug delivery: New approaches and *in vitro/in vivo* evaluation. *Int. J. Pharm., 235,* 1–15.

122. Kim, I. S., & Oh, I. J., (2005). Drug release from the enzyme-degradable and pH-sensitive hydrogel composed of glycidyl methacrylate dextran and poly(acrylic acid). *Arch. Pharm. Res., 28,* 981–987.

123. Rao, N. A., (1990). Role of oxygen free radicals in retinal damage associated with experimental uveitis. *Trans. Am. Ophthalmol. Soc., 88,* 791–850.

124. Nobuhiko, Y., Jun, N., Teruo, O., & Yasuhisa, S., (1993). Regulated release of drug microspheres from inflammation responsive degradable matrices of crosslinked hyaluronic acid. *J. Control. Release, 25,* 131–143.

125. Ullah, F., Othman, M. B. H., Javed, F., Ahmad, Z., & Akil, H. M., (2015). Classification, processing and application of hydrogels: A review. *Mater. Sci. Eng. C., 57,* 411–433.

126. Murakami, Y., & Maeda, M., (2005). DNA-responsive hydrogels that can shrink or swell. *Biomacromolecules, 6,* 2921–2929.

127. Yu, J., Fan, H., Huang, J., & Chen, J., (2011). Fabrication and evaluation of reduction-sensitive supramolecular hydrogel based on cyclodextrin/polymer inclusion for injectable drug-carrier application. *Soft Matter, 7,* 7381–7394.

128. Schmaljohann, D., (2006). Thermo- and pH-responsive polymers in drug delivery. *Adv. Drug Deliv. Rev., 58,* 15, 1651–1670.

129. Gupta, P., Vermani, K., & Garg, S., (2002). Hydrogels: from controlled release to pH-responsive drug delivery *Drug Discov. Today, 7,* 561–579.

130. Park, S. Y., & Bae, Y. H., (1999). Novel pH-sensitive polymers containing sulfonamide groups. *Macromol. Rapid Commun., 20*(5), 261–273.

131. Lee, Y. M., & Shim, J. K., (1997). Preparation of pH/temperature responsive polymer membrane by plasma polymerization and its riboflavin permeation. *Polymer, 38,* 1221–1232.

132. Soppimath, K. S., Kulkarni, A. R., & Aminabhavi, T. M., (2001). Chemically modified polyacrylamide-g-guar gum-based crosslinked anionic microgels as pH-sensitive drug delivery systems: Preparation and characterization. *J. Control. Release, 75*(3), 331–345.

# CHAPTER 3

# Application of Bioresponsive Polymers in Drug Delivery

MANISHA LALAN,[1] DEEPTI JANI,[1] PRATIKSHA TRIVEDI,[1] and DEEPA H. PATEL[2]

[1]*Department of Pharmaceutics, Babaria Institute of Pharmacy, NH#8, PO. Varnama, Distt. Vadodara–391247, Gujarat, India, E-mail: manisha_lalan79@yahoo.co.in (M. Lalan)*

[2]*Associate Professor, Department of Pharmaceutics, Parul Institute of Pharmacy and Research, Faculty of Pharmacy, Parul University, P.O. Limda, Ta. Waghodia, Vadodara–391760, Gujarat, India, Phone: 02668-260287, Fax: 02668-260201 E-mails: deepaben.patel@paruluniversity.ac.in, Pateldeepa18@yahoo.com*

## 3.1 INTRODUCTION

Few polymeric materials have the capability of responding to environmental stimuli to alter their physical and/or chemical properties reversibly/irreversibly. This change in response to the stimuli may be harnessed for achieving drug delivery goals. Such polymers are quintessentially termed as smart or intelligent or bioresponsive material. Such material may exhibit the dynamic change very quickly in the order of minutes and extending up to hours. A few of them show reversibility similar to biopolymers like proteins and polysaccharides while in others the change may be a permanent one. The alteration in physical or chemical property may be apparent as a change in individual chain dimension/size, shape, surface characteristics, three-dimensional structure, solubility, and degree of intermolecular bonding, etc. [1].

Bioresponsive polymers have displayed excellent utility in multifarious fields including drug delivery, diagnostics, and biosensing. The chapter delves into the intricacies and applications of such smart polymers in meeting drug delivery and targeting challenges. The use of polymeric material for maneuvering drug release from various dosage forms has a long history. But such systems have not exhibited sensitivity to physiological or pathological changes. This lack of control has inspired and driven the development of polymers which respond to the biological stimuli. The stimuli may vary in their nature from internal stimuli like pH, ionic strength, presence of metabolic chemicals, e.g., enzymes or antigens to external stimuli like magnetic fields, light, radiation, and ultrasound energy [2].

## 3.2  APPLICATION OF BIORESPONSIVE POLYMERS IN TOPICAL DRUG DELIVERY

The skin is the most formidable barrier against drug delivery but still exhibits typical characteristics which may be exploited in the design of bioresponsive systems. The pH of the skin surface, or stratum is typically in between the range 5.0–6.0 and termed as the 'acid mantle.' pH of the skin is influenced by number of internal and external factors, including epidermal cells, various secretory glands (like sebaceous, apocrine, and eccrine), age, gender, etc. [3].

Researchers have explored pH-sensitive hydrogels extensively in topical drug delivery applications. Suzuki and Tanaka worked on visible light-responsive hydrogels using the trisodium salt of copper chlorophyllin in PNIPAm hydrogels. The chromophore of the polymer absorbs the incident light to bring about a local increase in temperature. The temperature difference changes the swelling characteristics of the polymer and thus altering the viscoelastic behavior [4].

Hydroxyethyl cellulose-hyaluronic acid (HA) complex hydrogels have been investigated for the transdermal delivery of Isoliquiritigenin. The ratio of polymer in the complex was varied to obtain optimal rheological and adhesive properties, and was used to investigate the drug release efficiency as a function of pH. It was observed that the drug release efficiency was greater than 70% at pH 7. In addition, antimicrobial activity assays against *Propionibacterium acnes* were conducted to make use of the pH-sensitive properties of hydroxyethyl cellulose/hyaluronic acid (HECHA) [5].

Pathological specificities have also been exploited for the design of bioresponsive systems by incorporating disease-specific enzyme cleaved peptides. Aimetti et al. prepared a PEG hydrogel-based drug delivery system which imbibed human neutrophil elastase (HNE)-sensitive linkers for the management of inflammation. HNE is a serine protease secreted by neutrophils, which accumulates at sites of inflammation. They synthesized HNE-sensitive peptides using solid-phase FMOC chemistry and studied their degradation kinetics. It was observed that the rate of substrate degradation can be manipulated by the incorporation or substitution of specific amino acids. The utility of the bioresponsive system in local, controlled release from hydrogels containing HNE-sensitive peptides was visualized by fluorescence energy resonance transfer [6].

Wang et al. reported an anti-PD-1 loaded MN patch for sustained drug delivery which showed a glucose-dependent degradation for the melanoma treatment. The anti-PD-1 was crucial in blocking the programmed death-1 (PD-1) pathway. They were encapsulated in glucose-responsive nanocarriers and successfully showed immune responses B16F10 mouse melanoma model. The *in-vivo* response of the developed system was far better as compared to intratumoral injection or patches that were not triggered [7].

Enzymes which are activated in disease conditions or are overexpressed in specific disease can also serve as triggering mechanism for responsive systems. Hyaluronidase (HAase) enzyme is reported to be over-expressed in tumors and HAase activated system for immunotherapy for tumors was developed. 1-methyl-DL-tryptophan which inhibits immunosuppressive enzyme indoleamine 2,3-dioxygenase was conjugated with HA. This conjugate self-assembled into nanocapsule with anti-PD1 antibody as payload. These nanocarriers in combination with microneedle array are able to reach skin-resident dendritic cells (DCs) surrounding the melanoma tumor. The increased activity of HAase enzyme drove the drug release from the system [8].

Electro-responsive drug release from electro-conductive hydrogels was explored from semi-interpenetrating networks containing a blend of poly[ethyleneimine) and 1-vinylimidazole polymer as the novel electro-active species. The semi-interpenetrating networks are systems comprised of poly(acrylic acid) (PAA) and poly(vinyl alcohol). It was observed in the study that the relative ratios of the two polymers in the blend markedly influenced the degree of electro-responsive drug release and the matrix resilience of the hydrogels. The absence of an electric field resulted in low fluctuation flexibility and thus a highly stable molecular architecture. The use of an electric-stimulus offers versatility in usage of stimulus in terms of the duration of electric pulses, intervals between pulses, magnitude of current, etc. This gives an opportunity for precise control over the changes in physical structure. A number of polymers may be used as electroactive polymers like polypyrrole, polythiophene, polydimethylsiloxane, poly(methyl methacrylate), poly(3,4-ethylene dioxythiophene) and polyvinyl alcohol. These polymers have characteristic redox properties which render them semiconducting in nature and allow controlled ionic transport through the polymeric membrane. They allow for drug transport in the presence of an electric field only where they become resilient in nature and the drug release ceases in its absence. They may be used for continuous, pulsed, triggered drug delivery applications [9].

Thermoresponsive polymers can be of great utility in designing hydrogel systems for topical drug delivery where even self-triggering by the patient can be done. One of such studies was carried out on thermoresponsive polymer; poly(N-vinyl caprolactam) (PNVCL) based gel incorporating acetamidophenol and etoricoxib as model candidates. The transition temperature of the polymer was set to 35°C by grafting on to biopolymer chitosan (CS) to synthesize chitosan-g-PNVCL (CP) co-polymer. The synthesized copolymer exhibited pH sensitivity along with thermoresponsive characteristics. *In vitro* evaluations on rat abdominal skin demonstrated its utility as an on demand drug releasing system [10].

A thermoresponsive gel of Pluronic F127 was investigated for transdermal delivery indomethacin. Similarly, poloxamer 407 gel was investigated for insulin delivery. The drug delivery from smart polymers could be augmented with the use of supportive technologies like iontophoresis and use of chemical permeation enhancers [11].

Wound healing is one such area where smart drug delivery system can be a boon. Stimuli-responsive systems would be an approach which adapts to wound microenvironment and regulates the healing process by multiple triggers like thermal, chemical, biochemical, etc. One such prototype is that of nanofibrous mats prepared by electrospinning and blending to get an inflammatory mediators dependent matrix. A series of responsive HEMA and DMAEMA/HEMA scaffolds with a range of pH sensitivity were investigated for improved healing by promoting tissue formation. These scaffolds swelled in acidic microenvironment and permitted increased oxygen penetration and cell infiltration. The observations suggested improved wound healing with the scaffolds. Matsuda et al. also worked on photocross linkable gelatine to design a novel tissue adhesive technology. The developed system forms a gelatinous gel *in situ* on diseased tissue and gave sustained drug release. The photopolymerizable hydrogel systems could be exploited for local drug delivery in case of wound healing [12–14].

Microneedle arrays are the most researched transdermal drug delivery systems these days. The possibility of breaching the stratum corneum barrier with ease has opened plethora of opportunities for trancutaneous delivery of large molecules, hydrophilic moieties, vaccines, etc. Several variations of microneedle array have been investigated but hydrogel-forming microneedle assemblies are of prime interest as they can be tailored to give stimuli sensitive drug release. These hydrogel microneedles imbibe interstitial fluid to form swelled structures inside the skin which can give precise control over drug release. Donelly et al. designed microneedles which displayed triggered drug release in response to application of electricity. They used iontophoresis coupled with microneedles to investigate small molecules as well as peptide delivery. The studies yielded encouraging results for proteins and peptides. Another group of scientists have incorporated light-responsive drug conjugates in the hydrogel-forming microneedles to develop a system which would cater to on demand drug delivery challenges. Poly(2-hydroxyethyl methacrylate) (HEMA) was crosslinked by ethylene glycol dimethacrylate (EGDMA) and micro-molded to get sturdy microneedles. The system showed the drug release of ibuprofen for an extended 160 hours in response to light. The designed system was endowed with switch on and off characteristics because of the light-sensitive component.

Another interesting study by Ke et al. was on microneedles filled pH-responsive poly(lactic-co-glycolic acid) (PLGA) hollow microspheres to deliver multiple drugs simultaneously to skin. The hollow PLGA microspheres had an aqueous core containing model drug along with sodium bicarbonate. Another drug was loaded into the polyvinylpyrrolidone based microneedles. The polyvinyl pyrrolidone dissolved rapidly in skin on application releasing the drug while sodium bicarbonate generated carbon dioxide bubbles in the acidic microenvironment of skin. These bubbles created micropores in the PLGA shell to release the drug incorporated inside microspheres [15–18].

Apart of dissolving microneedles, even coated microneedles have been designed for bioresponsive delivery. A model antigen ovalbumin was coated on microneedles with pH-sensitive pyridine surface. The layer-by-layer assembly of polyelectrolytes gave reduced electrostatic interactions once inserted into acidic conditions of the skin. These interactions permitted the efficient release of ovalbumin [19].

Microneedles are designed with varying geometries and densities and can approach the dermal microcirculation also. Yu et al. designed a smart insulin patch, which released insulin in hyperglycaemic conditions. They prepared and incorporated vesicles from hypoxia sensitive HA incorporating insulin and GOx enzymes at the microneedle tips. The HA was conjugated with bioreducible 2-nitroimidazole group. When glucose is converted into gluconic acid it generated a hypoxic condition and it triggers the conversion of 2-nitroimidazole to hydrophilic 2-aminoimidazole and disassembly of the vesicles. The loss of structural integrity of vesicles brought about the release of insulin. The *in-vivo* studies in chemically induced type 1 diabetic mice showed the maintenance of euglycemic state within half an hour [20].

The research in bioresponsive transcutaneous systems has really paced up with innovative and hybrid approaches. Gu et al. have designed a hybrid system wherein a microneedle patch is associated with insulin-secreting beta cells along with glucose signal amplifiers. The microneedles were loaded with vesicles containing glucose amplifiers. The amplified glucose signal diffused into the microencapsulated beta cells located at the base of microneedle patch and triggering the secretion of insulin. *In-vivo* studies on the experimental system demonstrated maintenance of glucose levels for 6 hours in experimental animal model. Hydrogen peroxide produced in enzyme-catalyzed oxidation of glucose can also act as bio-trigger to mediate insulin release. $H_2O_2$-sensitive polymeric vesicles containing insulin were loaded into microneedle array patch. The building material for vesicles was a block polymer incorporated with polyethylene glycol and phenyl boronic ester conjugated polyserine. The block copolymer was degraded in presence of hydrogen peroxide and led to disassembly of polymeric vesicles. The insulin release from these vesicles could successfully control glucose levels in diabetic mouse model [21, 22].

Both of the earlier described strategies were combined to design hypoxia and $H_2O_2$ dual-sensitive vesicles loaded microneedles. The vesicles were prepared from diblock copolymer consisting of poly(ethylene glycol) and polyserine modified with 2-nitroimidazole via a thioether moiety. Bioreducible 2-nitroimidazole was converted into hydrophilic moiety under hypoxic conditions. Vesicles encapsulated both GOx and insulin where the oxygen consumption and hydrogen peroxide generation brought about a change in aqueous solubility of polymer. This led to dissociation of vesicles and subsequent release of insulin from them. The thioether moiety in this system offered additional advantage of scavenging hydrogen peroxide to eliminate the free radical-induced skin damage. The so formed vesicles were effective in maintaining euglycemia for more than 10 hours in diabetic animal model [23].

Zhang et al. developed a thrombin responsive system for heparin delivery in the management of coagulation disorders. Heparin was coupled to HA through a thrombin-cleavage peptide. The microneedles released heparin in response to increased thrombin from the blood clots. Such a system promises to be utility in the management of abnormal blood clotting and acute pulmonary thromboembolism [24].

The number of biopharmaceuticals and biotechnology-based products in treatment and management of disorders are increasing every day. A new transcutanous approach towards delivery of polyplex based DNA vaccines utilizes microneedles coated with pH-responsive polyelectrolyte multilayer assembly. Such a system not only maximized the polyplex release, but was coupled with other strategies to surmount a sustainable and prominent humoral immune response [25].

## 3.3 APPLICATION OF BIORESPONSIVE POLYMERS IN ORAL DRUG DELIVERY

Oral drug delivery is the most sought after route of delivery but often a challenge for the formulation scientist. There is gamut of changes across the entire length of GIT with respect to pH, ionic profile, enzymes, microbial load, etc. Although these variations may be a constraint in designing oral dosage form, they also offer opportunities for on demand, triggered, and targeted drug release. Also, different diseases are associated with hallmark variations in physiological parameters like malignancies, autoimmune disorders, degenerative disorders, infections, and cardiovascular pathologies, which make them targets when designing bioresponsive systems.

A large number of polymers have been reported for bioresponsive drug delivery applications. These stimuli-responsive polymers have also been combined or conjugated with several different types of bioactive molecules like enzymes, antibodies, peptides, nucleic acids, organic molecules such as steroids, anticoagulants, heparin, HA, etc. Such polymers have also been conjugated with hydrophilic moieties like, poly(ethylene glycol) to impart stealth properties.

pH-responsive oral delivery has a long history of more than 65 years. Enteric-coated dosage forms were the first of the responsive systems made available to clinics. A large number of polymers have been explored but majority of the pH-sensitive polymers belong to one of the two groups. The first group of polymers are basically copolymers of pH-sensitive methacrylic monomers while the second group is of cellulosic polymer where phthalic anhydride has been used to esterify some of the -CH$_2$OH groups of the backbone. The carboxyl groups of both the groups of polymers are protonated and unionized at gastric pH and hence, are hydrophobic in nature. They become hydrophilic at intestinal pH and get ionized and thus permitting drug release of acid-labile drugs in a more favorable intestinal pH. Basically anionic polymers are used for enteric protection and colon-specific drug delivery and improving the bioavailability of weakly basic drugs.

The pH sensitivity of polymers has also been exploited for taste masking. The pH difference between the buccal cavity (pH 5.8–7.4) and the stomach (pH 1–3.5) is commonly used by pH-responsive polymers to control drug release and impart taste-masking characteristics. Amine group bearing cationic polymers possess higher water solubility at acidic pH than at neutral pH. These polymers with pH-dependent dissolution characteristics are widely employed for masking the taste of bitter drugs.

Aminoalkyl methacrylate copolymer (Eudragit E ®) is a well known safe and approved cationic polymer which exhibits high solubility below pH 5. Researchers have explored the Eudragit E coated microspheres in delivering bitter drugs like sumatriptan succinate and donepezil hydrochloride. The microspheres which were prepared by spray drying technique could limit the drug release in neutral pH or simulated salivary fluid thus, masking the bitter taste of drugs. The developed formulations showed fast and immediate release in gastric pH and were similar to marketed products. Such similar studies are available on a number of drug candidates including atorvastatin and promethazine [26, 27].

Poly(methacrylic acid-co-methyl methacrylate) (Eudragit L, S, and F), hydroxypropylmethylcellulose phthalate and HPMC acetate succinate, they have carboxyl groups on the polymer side chains. These polymers are insoluble at low gastric pH but soluble at neutral intestinal pH. There are countless research studies and even significant number of products in the market which are based on such polymers. They are very apt for acid-labile products and offer a cost-effective solution to drug delivery issues. The proton pump inhibitors are a group of drugs which is well known for its acid sensitivity have been extensively investigated for such applications. A number of formulations like enteric-coated tablets, capsules, beads, granules, etc., have been developed. Not even small molecules but large molecules like bovine serum albumin (BSA) has also been encapsulated in such enteric-coated microspheres which displayed negligible drug release at pH 1.0 but ensure unimpeded release at pH 6.8.

Another major application of ph responsive polymer is colon-specific drug delivery. The small intestine is a site for enzymatic degradation of proteins and peptides as it harbors a dense bacterial population. Hence, colon is a preferred site for delivery of proteins and peptides and other enzyme labile drugs. Eudragit F is one example of enteric polymers used for colon-targeted DDS as it has solubility at pH higher than 7.0. Though a lot of drugs have been researched for colon-specific drug delivery but number of drugs, reaching the market is quite less. One of the prototype examples is that of Eudragit S100 coated citrus pectin nanoparticles, which were formulated for the delivery of 5-Fluorouracil specifically in colon [28].

The management of local diseases of gastrointestinal tract like inflammatory bowel disease may become more efficient if a bioresponsive system releases drug at the site of inflammation. Zhang et al. developed an inflammation targeting hydrogel microfibers of ascorbyl palmitate which were loaded with model drug dexamethasone, it adhered to the inflamed mucosa and released drug at the site by enzymatic digestion. The novel system displayed success in both *in-vitro* and *in-vivo* conditions [29].

One of the platform technologies in pH-responsive systems is based on hydrogels comprised of itaconic acid copolymerized with N-vinylpyrrolidone. They were explored for the delivery of proteins exhibiting high isoelectric point. The study was carried out using salmon calcitonin and results were quite encouraging as the system showed intestinal release with faster and more extensive pH-responsive swelling than methacrylic acid-based hydrogels. Another investigation by the same group on enzymatically-degradable hydrogels of P(MAA-co-NVP) crosslinked with the peptide sequence MMRRRKK also yielded encouraging results for selective salmon calcitonin release in intestinal condition [30].

A copolymer of NIPAAm and acrylic acid was developed which had both pH and temperature sensitivity for use in the preparation of matrix drug delivery system. The drug was admixed physically with the developed copolymer. Although this was, a matrix system but it behaved very similar to enteric coating. The system remained intact and insoluble at acidic pH and gradually dissolving at intestinal pH. The developed formulation did not release rapidly in intestine in contrast to enteric-coated tablets, but gave sustained release of drug over an extended period of time. The rate of drug release directly correlated with the amount of acrylic acid [31].

A number of stimuli-responsive insulin delivery systems have been investigated. Morishita et al. prepared microparticles with poly(methacrylic acid-g-poly(ethylene glycol)) P(MAA-g-EG) which gelled in an acidic environment to delay the insulin delivery in neutral or alkaline pH. Foss et al. formulated poly(AA)-g-PEG nanoparticles for oral insulin delivery. The nanoparticles exhibited unique characteristics. The size of the nanoparticles became more than 600 nm at pH 6 which would avoid uptake by intestinal Payer's patch [32].

Insulin is conventionally delivered through nonpatient friendly subcutaneous route only. A new nanoparticulate system composed of CS and poly($\gamma$-glutamic acid) for oral delivery of insulin was designed. The developed system exhibits pH responsiveness and CS which adheres to mucosal surfaces and brings about transient opening of tight junctions. The intestinal pH leads to disassembly of nanoparticles and mucosal adherence of CS. The opening of junctions helps to deliver insulin paracellular [33].

Wilson et al. prepared a thioketal nanoparticle-based delivery vehicle for delivery of small interfering RNA (siRNA). It was a redox responsive polymer system for management in intestinal inflammation. Thioketal nanoparticles were prepared by poly(1,4-phenyleneacetone dimethylene thioketal), which degraded selectively in response to ROS. This will trigger the siRNA release under the high levels ROS of intestinal inflammation after oral delivery. The effectiveness of the delivery system was demonstrated by the suppression of TNF messenger RNA levels in the colon and prevention from ulcerative colitis in mice model [34]).

Not only synthetic polymers but even natural polymers can exhibit stimuli responsiveness. A polysaccharide (rhamnogalacturonan) based hydrogel from linseeds (*Linum usitatissimum L.*) showed swelling–deswelling behavior in response to pH. The hydrogel deswelled in acidic pH but swelled in neutral pH and pH-controlled drug release was observed [35].

Stimuli-responsive nanocarriers (SRNs) are unique nanosized delivery vectors which are endowed with "load and release" modalities in their constituent units. Any specific intracellular, extracellular physical, chemical, or biochemical stimulus will alter the structural properties of the nanocarriers and leading to change in drug release pattern. The dynamic changes observed are generally due to decomposition, isomerization, polymerization, supramolecular assembly, etc. The specificity w.r.t. pH, enzyme, protein over-expression, ionic changes allowed the nanocarriers to release their payload with spatio-temporal specificity. The precise drug release would minimize the adverse reaction and side effects. These carriers are actually simulating the feedback mechanism operating in nature where the presence, absence, or excess of any physical, chemical, or physico-chemical factors regulates a series of biochemical processes.

We have already discussed that polymers with ionizable groups, such as amines and carboxylic acids, are the ideal choice for fabricating pH-sensitive carriers. A PMAA-PMA-copolymer was used to enhance the oral bioavailability of cyclosporine A and release it at a pH of more than 6. The nanocarriers protect the drug moiety from acidic degradation and allowed drug release in favorable environs only [36].

A pH-responsive micelle was designed for oral application using poly(ethylene glycol)-b-poly(alkyl acrylate-co-methacrylic acid) (PEG-b-P(AlA-co-MAA)). The micelle core which was hydrophobic becomes hydrophilic on deprotonation. The deprotonation also destabilizes the assembly [37].

The same system could be modified with the use of acid degradable linker unit also. Kataoka and group worked on poly(ethylene glycol)-block-poly(aspartate) polymer coupling it with doxorubicin (DOX) by a pH labile hydrazone bond. The system was tagged with folate residues for additional tumor specificity. They also used the same strategy for PEG-b-PLLA micelle to deliver cisplatin in a pH-dependent fashion [38].

Yet another study investigated an interesting but different approach for responsive systems. A porous anionic metal-organic framework, bio-MOF-1, was designed with adenine and local anesthetic, procainamide hydrochloride was loaded into the pores of bio-MOF-1 using a simple cation exchange process. The system showed the release of adsorbed drug which was manipulated by exogenous cations from biological buffers [39].

Self-assembled cationic nanocomplexes were fabricated from interaction of poly(amidoamine)s with oppositely charged proteins. Two variants of poly(amidoamine)s were synthesized for the studies. These water-soluble polymers condensed human serum albumin (HSA) by self-assembly into stable nanoscaled and positively-charged complexes. They had mucoadhesive properties and rapidly destabilized in intracellular compartment because of cleavage of the repetitive disulfide linkages. The nanocomplexes underwent significant cellular uptake in cell line studies [40].

Discher and Yang explored poly(ethylene oxide)-*block*-poly(*N*-isopropyl acrylamide) (PEO-*b*-pNIPAm) block copolymers to generate micelles which were temperature responsive. The amphiphilic characteristics of the polymer at above body temperatures permitted the encapsulation of both hydrophilic and hydrophobic molecules. The assembly was dismantled at body temperature and released the encapsulated small molecules [41].

Glucagon-like peptide-1 (GLP-1) is an incretin peptide of the endocrine L-cells of the intestinal mucosa with unique antidiabetic potential. Although the molecule has tremendous potential, its low absorption efficiency and instability in the gastrointestinal tract becomes a challenge for its oral delivery. A novel silica-based pH-sensitive nanomatrix of GLP-1 was developed which was composed of silica nanoparticles and pH-sensitive Eudragit®. The cell line studies, pharmacokinetics, and intraperitoneal glucose tolerance test, toxicological evaluations validated its success. Another study on mesoporous silica nanoparticles (MSPs) with surface positive charges could entrap anionic molecules. The entrapped drugs were not released in acidic pH and released only in the neutral intestinal pH [42, 43].

Externally administered small molecules have also been used as stimuli to trigger drug release. For example, orally administered antibiotics have been used to destabilize drug delivery vehicles which have antibiotic-sensitive units to trigger drug release. A hydrogel composed of the bacterial gyrase subunit B (GyrB) coupled to polyacrylamide was formed by dimerization of the GyrB by the antibiotic coumermycin. Administration of novobiocin competitively displaced coumermycin and leading to hydrogel dissolution. The novel platform was used to for immunizing mice with oral hepatitis B vaccine [44].

Smart polymers which are sensitive to the change in temperature and modify their microstructural features are the most commonly used and most safe polymers in drug administration systems. Thermo-responsive polymers exhibit these temperature-sensitive characteristics by striking a very sensitive balance between the hydrophobic and the hydrophilic groups. A minor change in the temperature can create new adjustments and alter structural conformations. The most common monomers are N-alkyl substituted poly(acrylamides), especially the poly(N-isopropyl acrylamide) (PNIPAAM) which undergoes a sharp phase transition at 32°C is the most widely explored polymer. The structural alterations are driven by entropic effects. The interest in PNIPAAM for drug delivery is by virtue of its safety profile and LCST close to body temperature. The transition temperature can be further adjusted by varying the alkyl part or upon copolymerization with other monomers [45].

Apart from internal stimuli like pH and temperature, external stimuli like magnetic field and ultrasound energy can also be exploited for achieving targeted drug release. In one of the studies, acyclovir depot tablets with internal magnets were developed. Extracorporal magnet was used to extend the duration of residence in gastric media. Clinical studies were carried out and pharmacokinetics parameters confirmed the prolonged retention of tablets [46].

Another study explored ultrasound controlled drug release platform for diclofenac-loaded alginate microcapsules (fabricated with a home-made electrostatic device, 75% embedded rate). The study established the anti-inflammatory efficiency in the triggered environment. Both continuous and pulsatile irradiation was investigated and the results indicated superior performance with pulsatile irradiation [47].

## 3.4 APPLICATIONS OF NASAL DRUG DELIVERY SYSTEMS

Nasal drug delivery has attracted renewed and significant research interest in the last years because of the possibility of multi-variant targeting. Nasal routes yields to local drug delivery, systemic drug delivery, as well as brain drug targeting. The advantages of self-administration, noninvasive and bypass of hepatic first-pass metabolism make it an attractive and potential route of drug delivery. It is one of the alternative routes of delivery which becomes more relevant and important for biopharmaceuticals. Intelligent drug delivery systems which respond to physiological or eternal stimuli have emerged as another game-winning strategy for formulation scientists. The phase transitions triggered by biological or external stimuli postnasal administration ascertains sufficient contact time for drug delivery, better permeation profiles and thus improved bioavailability [48–50].

The interest in *in-situ* nasal systems is because of the drawbacks of earlier dosage forms like nasal sprays and nasal drops. The low viscosity of the systems leads to their early clearance while higher viscosity made it impossible to administer or spray in the nasal cavity. Thus, the smart polymer-based systems became relevant and enticing. The systems are low viscosity and easily administered and will undergo phase transitions *in-vivo*. The nasal applications of such smart polymers are generally confined to three approaches namely, thermoresponsive, pH-responsive, Ion responsive systems. Thermo-responsive systems which have been investigated are usually mucoadhesive formulations of polymers which exhibit temperature catalyzed sol-to gel transitions in the range of 25–37°C. The temperatures below the lower limit and above body temperature precipitate problems like early or late gelation, which will impact the ease of handling or lead to liquid formulation's leakage and early clearance from the nasal cavity [51].

Literature reveals a large number of research studies on thermoresponsive gels for nasal administration. We will present a bird's eye view on the subject. Poloxamers have been one of the most researched thermoresponsive polymers. In one of the study, Majithiya et al. developed a composite gel based on Poloxamer 407 and Carbopol 934P for sumatriptan succinate in management of migraine. The transition temperatures were below 30°C. The gel showed mucoadhesive characteristics and enhanced permeability and induced no cellular toxicity [52]. Zaki et al. also worked on Polaxamer 407 but with polyethylene glycol and multiple mucoadhesive polymers for metoclopromide. The inclusion of mucoadhesive polymers modified the rheological properties and increased the residence time as evidenced by longer mucociliary transport time. The formulation also ensured a faster Tmax compared to oral solution [53]. Ketorolac tromethamine nasal sprays gave insufficient nasal residence time to limit their applicability. Another study on thermoresponsive hydrogel of Poloxamer 407 and Carrageenan for ketoraolac ensured higher nasal retention *in vivo* studies (54).

Smart polymer-based hydrogels increased the residence time, but they may be exploited with other aims also. Rizatriptan benzoate used in the treatment of migraine suffers from extensive hepatic metabolism upon oral administration. Hence, Kempwade and Taranalli explored nasal delivery as an alternate to escape the first-pass metabolism. Thermo-reversible gels comprised of Poloxamer 407 and Carbopol were developed and investigated. Superior permeation and no toxicity of the developed formulation were observed [55].

Qian et al. formulated tacrine in a Pluronic F-127 (20%, w/v) *in situ* gel for the intranasal delivery. The formulation was rationalized by its low oral bioavailability (17–24%) and dose-dependent hepatotoxicity. *In vivo* studies revealed 2–3 fold higher peak plasma concentration (Cmax), a significantly higher tacrine exposure in the brain and reduction in drug's metabolites after intranasal administration of the gel, compared to the oral solution [56].

It is practically impossible for the hydrophilic molecules to gain entry into the brain by crossing blood-brain barrier. One of the approaches for direct brain targeting is through nasal cavity. The nasal cavity offers direct transport of drugs to brain through olfactory and trigeminal nerve cells. Gabal et al. loaded anti-Parkinsonism drug, ropinirole into nanostructured lipid carriers and integrated them in Poloxamer 188 *in situ* gels. The system vastly improved the bioavailability by direct nose to brain transport [57]. Another study on thermoresposnsive polymers for nose-to-brain delivery was for 32P-siRNA dendriplexes [58]. Poloxamer 407 along with muco-adhesives, carbopol, and CS were employed in formulation. Radiological studies confirmed the delivery of cargo in brain and showed better brain localization than intravenous or intranasal buffered administration. Another study in the same direction by Jose et al. explored Pluronic based thermo-sensitive hydrogels containing CS microspheres for direct nose to brain transport of lorazepam. The designed system gave sustained release of drug and was biocompatible also [59].

Breaching blood-brain barrier is extremely difficult for the hydrophilic molecules. Rationalized by the need of direct brain deposition of hydro-philic antidepressant agent venlafaxine hydrochloride, Bhandwalkar et al. developed thermorepsonsive mucoadhesive gel of Carbopol 934P, HPMC K4M, PVP K30, sodium alginate, Tamarind seed gum, and Carrageenan. The transition temperatures were very close to body temperature and proved to be effective in comparison to its oral congener [60].

Loratadine-b-CD (beta cyclodextrin) complexes were incorporated into Poloxamer 407 and Carbopol 934P nasal gels, in an attempt to spare the drug from rapid first-pass hepatic metabolism and circumvent its low oral bioavailability. *In situ* gels had retarded mucociliary clearance rate due to presence of carbopol. Drug transport evaluation in *ex vivo* studies revealed that more than 90% of the drug could be permeated at the time scale of 6 h, without affecting epithelium physiology [61].

Not only synthetic but natural polymers have also been used in designing stimuli responsive nasal drug delivery systems. Basu et al. [62] assessed the efficiency of natural mucoadhesive polymer extracted from F. carica, compared to synthetic congeners (hydroxypropylmethyl cellulose and Carbopol 934) in delivery of midazolam. Xyloglucans is one such group of thermoresponsive biopolymers which can be exploited in formulating *in situ* gels. Studies by Mahajan et al. investigated the utility of xyloglucans in delivery of ondansetron hydrochloride. Such systems increase the residence time and permit near complete drug absorption as was observed in *ex vivo* studies [63]. CS is a natural polymer obtained by N-deacetylation of chitin was integrated into a thermo-reversible gel of Poloxamer 407, hydroxypropyl-b-cyclodextrin (Hp-b-CD) and fexofenadine hydrochloride. Drug permeation enhancement was observed in cell monolyers and it translated to 18 fold higher bioavailability in rabbit model [64, 65].

Derivatives of natural polymers have also exhibited stimuli responsive behavior. Tri methyl CS displayed thermosensitive behavior when co formulated with polyethylene glycol and glycerophosphate. Such formulations when loaded with insulin and their subsequent nasal application demonstrated reduction in blood glucose over a time span of 24 hours. The cell lines studies evidenced that there is reduction in transepithelial resistance, opening of tight junctions to permit paracellular transport of insulin. The *in situ* gels showed extended retention in nasal cavity by virtue of their viscosity and bioadhesive characteristics [66, 67]. Another study assessed a thermosensitive hydrogel comprised of N-[(2-hydroxy-3-trimethylammonium) propyl] CS chloride, poly(ethylene glycol), a-b-glycerophosphate, and insulin for hypoglycaemic effect. The study results were in agreement of previous such study with marked plasma glucose level reductions [68].

Ion responsive systems are another approach for designing *in situ* gels by using ion responsive agents. Gellan gum, an anionic polysaccharide, undergoes gelation in presence physiological cations. Gelling occurs due to double helical junction zones formation and inter-helical interaction, generating a three-dimensional network via cations complexation. The presence of cations, especially $Ca^{2+}$, in nasal fluids, can catalyze *in situ* gelation of gellan gum formulations in nasal cavity to sustain their residence and improve drug absorption [69]. Cao et al. explored gellan gum formulations for trans-nasal transport of an anti-muscarinic agent. The same group also investigated further in the same line to improve the nasal absorption efficacy of mometasone furoate. The gellan gum-based formulations were safe for nasal administration and displayed good formulation stability [70, 71].

Galgatte et al. employed deacetylated gellan gum as the ion-responsive gelling agent for trans nasal delivery of sumatriptan succinate to by-pass the extensive hepatic metabolism of the drug. The studies indicated towards a direct nose to brain transport of the drug [72]. Wang et al. also worked on deacetylated gellan gum in design of curcumin loaded microemulsion based ion responsive hydrogels. Bioavailability and brain targeting index were improved markedly [73].

Nanocarriers incorporated in such stimuli-responsive carriers will offer added advantages of improved permeation along with bioresponsive behavior. A nanosuspension of carvedilol was formulated as an *in situ* gelling nasal spray containing gellan gum. The success was validated in an animal model where the novel system demonstrated a 2.7-fold higher bioavailability [74].

Pectin is also one of the natural polymers which can gel in nasal fluids. Castile et al. investigated the nasal deposition and gelling properties of PecSys, low methoxy pectin. The clinical utility of pectin *in situ* gel has been evaluated in trials for the trans-nasal transport of fentanyl, and was similar to the findings of the *in vitro* models developed by Castile et al. [75]. Pectin extracted from Aloe Vera plant was evaluated as an *in situ* gelling agent. It was an effective gelling agent at low concentrations of calcium. Advanced studies characterized the gelation process in detail and proved the utility of the system [76, 77].

Another approach utilizes freeze drying to develop *in situ* gelling bioadhesive nasal inserts. The process of lyophilization created porous structures to ensure water penetration for gel formation. Using oxymetazoline as the model drug, Carrageenan, xanthan gum, and carboxymethylcellulose sodium were useful matrix components of the nasal inserts. The same group of researchers further evaluated the potential of the system by designing sponge-like nasal inserts for influenza vaccine delivery [78–80]. Xanthan gum was used as the matrix component of the inserts and cationic lipid, as adjuvant, could mount a satisfactory immune response, reflected in the IgG levels. Another nasal insert was introduced by Farid et al. with a number of polymers like sodium alginate, CS, hydroxypropylmethylcellulose, and carboxymethylcellulose sodium-containing salbutamol sulfate as the model drug. The polymers gave significant mucoadhesion and favorable drug release [81].

Nakamara et al. developed microparticles from bioadhesive graft copolymers of polymethacrylic acid and polyethylene glycol (P(MAA-g-EG)).

Budesonide was loaded into the microspheres from ethanolic solutions and the system showed pH-responsive behavior. The drug release from swollen microparticles was fickian after an initial burst. The pharmaco-kinetics showed a peak concentration of the drug approximately 45 min after administration [82]. Nasal drug delivery systems of polyvinylacetal diethylaminoacetate (AEA) were formulated to evaluate drug release *in vitro* and *in vivo*. The hydrogel formation was observed when the formulation was instilled in nasal cavity. The drug release was biphasic with an initial rapid release followed by sustained release as the hydrogel formation takes place [83].

CS is another polymer which is known to exhibit pH-responsive characteristics. This was utilized in designing nanocarriers for siRNA delivery. Imidazole-modified CS (CS-imidazole-4-acetic acid [IAA])-siRNA nanoparticles were formulated to mediate gene silencing after intranasal administration. An effective silencing of glyceraldehyde 3-phosphate dehydrogenase (GAPDH) protein expression was seen in the lungs at a dose of 0.5 mg/kg/day siRNA delivered over three consecutive days. The studies put forward the use of pH-responsive nanocarriers in nasal drug delivery of biopharmaceuticals [84].

Another study explored the surface-functionalized, pH-responsive poly(lactic-co-glycolic acid) (PLGA) microparticles for nasal delivery of biopharmaceuticals, hepatitis B surface Antigen. Double emulsion technique was employed to prepare pH-responsive PLGA, chitosan modified PLGA (CS-PLGA), mannan modified PLGA (MN-PLGA), mannan and chitosan co-modified PLGA (MN-CS-PLGA) microparticles. Antigen release was rapid between pH 5.0–6.0 and was hindered at pH 7.4. The antigen release pattern was varied in the different variants of microparticles. PLGA and MN-PLGA microparticles showed rapid release in endosomes/lysosomes. Slow and gradual release from CS-PLGA microparticles in cytoplasm and a combination of fast release and slow release patterns from MN-CS-PLGA microparticles. *In vivo* immunogenicity studies indicated stronger humoral and cell-mediated immune responses with MN-CS-PLGA microparticles in comparison to others. The surface modifications of pH-responsive PLGA microparticles further augmented their utility [85].

Core shell-type nanoparticles were prepared with a poly($\beta$-amino ester) (PBAE) core and phospholipid bilayer shell for mRNA delivery. The pH-responsive characteristics were imparted by the PBAE component while the lipid surfaces help in minimizing toxicity of polycation core. mRNA loading was facilitated by the electrostatic interactions. *In vitro* testing rationalized the hypothesis with mRNA delivery in cytosol and effective transfection. Intranasal administration in rats verified the transfection success with reporter protein luciferase expression in 6 h. The transfection efficiency was significantly higher than naked mRNA delivered intranasal [86].

Magnetophoretic-guided delivery has been shown to improve the drug transport across nasl mucosa. This strategy is challenging because of the complex anatomical structure and rapid decay of magnetic intensity which hampers the control of particle motion. Optimization of the guiding system can help it adapt to intranasal drug delivery [87].

Rationalized by the need for developing newer methods to enhance drug delivery to olfactory region, Xi et al. investigated magnetophoresis to improve olfactory delivery efficiency. The study revealed the effect of magnet layout, magnet strength, drug-release position, and particle diameter on the olfactory dosage. It was observed that particles sized around 15 μm could be effectively guided by magnets and optimal combination of magnet layout, selective drug release, and microsphere-carrier diameter could help in improving transnasal drug delivery. It was predicted the magnetophoresis can increase drug delivery up to 64 folds. However, optimizing the operating conditions and unstable nature of the magnetophoresis is still a challenge [88].

Focused ultrasound can be used as other stimuli guided approach for drug delivery. Brain-derived neurotrophic factor (BDNF) was subjected to focused ultrasound following nasal delivery. Immunohistochemistry staining of BDNF showed that focused ultrasound enhanced drug transport by active pumping. It was hailed as a promising technique for noninvasive and localized drug delivery [89].

## 3.5 APPLICATIONS OF STIMULI-RESPONSIVE POLYMERS IN OCULAR DRUG DELIVERY

As vision is an important sense in humans, stimuli-responsive polymers are gaining interest of researchers in development of ocular drug delivery system. There is increase in attention on ocular disorders/diseases due to the increase in the aging population worldwide. Stimuli-responsive poly-mers have great potential in optimizing and improving the efficiency of optical disorder management strategies. These systems have the capability of controlling drug delivery in response to various stimuli such as pH, temperature, light, and change in level of biomarkers. Stimuli-responsive polymers offers great benefits over traditional systems since release of drugs can be controlled, resulting in more specific effect and less side effects [90, 91].

Various types of stimuli-responsive ocular drug delivery system are: Physical stimuli: light, temperature, electric fields, magnetic field Chemical stimuli: ion activated, pH micro electro mechanical system (MEMS) actuate in response to various stimuli including temperature, electric field, magnetic field or pressure to carry out their function. They made up of one or more drug reservoirs containing either a single drug or multiple drugs in different reservoirs and actuators to release drug in response to the stimulus. Control over drug release due to this system, result in accurate drug delivery. The MEMS-based MicroPump™ System for clinical use is tested for the treatment of various ocular conditions with a lifetime of up to five years. This system can be used in future, to release drug in response to increased intraocular pressure in the glaucoma [90].

Pirmoradi et al. studied magnetic stimuli-responsive drug release. Magnetically responsive reservoir systems included magnetically responsive components to control drug release. One such implantable device containing docetaxel (DTX) has been investigated for the purpose of drug delivery to the posterior segment of the eye in the treatment of proliferative vitreoretinopathy. A 64-fold increase in drug release was noted in the presence of a magnetic field (213 mT magnetic field strength). This device allowed control over drug release depending on the strength and duration of the magnetic field applied [90].

The ocular route possesses a great potential for opthalmologically-active therapeutic peptides and proteins for use in the treatment of ocular diseases.

Light-activated drug delivery systems, to treat ocular disorders, have been prepared by incorporating light-sensitive materials into the formulation. The light-sensitive materials are activated either in response to a specific wavelength or by utilizing heat generated from light.

Electromagnetic radiations (EMR) like ultraviolet (UV) ranging from 250–380 nm and near-infrared (NIR) ranging from 700–900 nm are generally used to activate these systems. Unionized leuco-derivatives are generally used in UV sensitive drug delivery systems. Under the influence of UV radiation, these compounds are ionized and created an osmotic pressure in the polymer/ gel system, which causes solvent influx, resulting in increased drug release. Micelles having chromophore binding on the hydrophobic part of the polymer are used generally in NIR sensitive drug delivery systems. Specific wavelength of light results into cleavage of chromophore-polymer binding and drug is released. Chromophore-polymer binding can be reversible or irreversible affecting drug release for short or extended periods of time [92–96].

Photodynamic therapy (PDT) by porphyrins and related tetrapyrrole derivatives is treatment modality of age-related macular degeneration (AMD). This therapy combines the administration of porphyrins or porphyrin precursors and use of red light at the diseased sites.

Therapy of AMD is performed by giving i.v. injection of verteporfin (Visudyne) a hydrophobic porphyrin followed by red laser light, which leads to accumulation of the porphyrin in endothelial cells of choroidal neo-vessels [97].

Temperature responsive materials are effective for treatment of various retinal disorders. Verestiuc et al. developed a system for the delivery of pilocarpine hydrochloride and other ocular drugs. They used acrylic acid-functionalized CS, copolymerized with either *N*-isopropylacrylamide or 2-hydroxyethyl methacrylate monomers. Comparative *in-vitro* studies performed using chloramphenicol, atropine, norfloxacin, or pilocarpine indicated that for the controlled delivery of these compounds, selection of drug-specific carrier compositions is needed. In addition, *in-vivo* studies of pilocarpine in rabbit model indicated use of CS-based hybrid polymer networks containing 2-hydroxyethyl methacrylate for the delivery of this therapeutic agent [98].

Von Recum et al. developed a thermoresponsive poly(N-isopropyl acrylamide)-based scaffold with degradability and controlled porosity. Effect was studied on the enzymatic integrity of donor retinal cells. They observed that the temperature sensitive materials do not produce ill effects on the retinoid enzymatic profile. They also found that these agents are least deleterious for cell growth and detachment [99, 100].

Derwent and Mieler studied the application of thermoresponsive hydrogel as drug delivery system for the treatment of AMD. AMD is the leading cause of blindness for the elderly. They reported use of vascular endothelial growth factor (VEGF) inhibitors at posterior segment of the eye as drug delivery site for the treatment of AMD. They synthesized thermoresponsive hydrogel using poly(N-isopropylacrylamide) cross-linked with poly(ethylene glycol) diacrylate. They found that this hydrogel is able to encapsulate and release various proteins including BSA, immunoglobulin G (IgG), bevacizumab, and ranibiumab [101, 102].

Xie et al. studied application of injectable thermosensitive polymeric hydrogel for the treatment of posterior segment disorders. They reported the use of injectable thermosensitive hydrogel to deliver Avastin® (bevacizumab) in the treatment posterior segment diseases. They developed Avastin(®)/PLGA-PEG-PLGA hydrogels, facilitating sustained release of Avastin® over a period of up to two weeks *in vitro*. They found that the PLGA-PEG-PLGA hydrogel exhibited no sign of retinal toxicity, which supports its use in ocular disorders [103].

Ninawe and colleagues investigated ocular delivery of proteins by using poly(N-isopolyacrylamide) (NIPAM)-based implantable hydrogel. They loaded a model protein immunoglobulin G (IgG) into the gel. They implanted formulation in the sclera for delivery to the posterior segment of the eye, to treat AMD. After reaching at body temperature, the gel starts to Deswell, resulting into drug release. Released drug diffuses across the sclera to reach at target site i.e., the choroid and pigmented layer of retina. When deswelling stops, the protein is released via diffusion. The hydrogel system reduced the frequency of dose administration as compared to intra-vitreal injection. This system shows increased selectivity for target tissue to protect normal tissue from damage by the therapeutic peptides [104].

Tsai et al. formulated electro-conductive hydrogels based on poly(vinyl alcohol), crosslinked with diethyl acetamidomalonate. They used polyaniline as the inherently conductive component, and fabricated in the form of cylindrical devices to confer electro-actuable release of indomethacin. The hydrogels were characterized for their physicochemical and mechanical properties. Drug entrapment efficiency was more than 60%. Drug release was studied by periodically applying-removing-reapplying an electric potential ranging from 0.3–5.0 V for 60 seconds at an hourly interval. The cumulative drug release obtained ranged from 4.7–25.2%. The electro-stimulated release of indomethacin was depended on the degree of crosslinking, the polymeric ratio, and drug content. A Box-Behnken experimental model was constructed employing 1.2 V as the baseline potential difference. This developed novel device exhibited superior swellability and high diffusivity of indomethacin, in addition to electrical switching induced "ON-OFF" drug release kinetics. In order to study the electro-actuable release of indomethacin, molecular mechanics simulations using AMBER-force field were conducted on systems containing water molecules and the poly(vinyl alcohol)-polyaniline composite under the influence of an external electric field. Various interaction energies were observed to study the effect of the external electric field on the erosion of polyaniline from the co-polymeric matrix. This approach allows for electro-actuable drug release from electro-conductive hydrogels in controlled and local drug release while sustaining a mild operating environment [105].

Degradation of proteins and peptides is major hurdle for use of these agents. pH-responsive systems can be used by ocular route to avoid degradation of proteins and peptides. pH of the tear fluid can be targeted for same (91).

Sultana and co-workers studied an ion-activated; gelrite based ophthalmic gels of pefloxacin mesylate. They found, this system effective in management of conjunctivitis [106].

Vodithala et al. developed ion-activated ocular *in situ* gels of ketorolac tromethamine. They used gelrite as a polymer. The formulation studied, showed sustained release of drugs for up to 6 h. They found that formulations are non-irritating. No ocular damage was observed [107].

Hui & Robinson reported the use of acrylates for ocular delivery of progesterone, based on viscosifying as well as bioadhesion properties. Carbomer (Carbopol) a cross-linked acrylic acid polymer showed pH-induced phase transition when the pH raised above its pKa of about 5.5 [108]. Rozier et al. reported an improvement in the ocular absorption of timolol administered with gelrite in albino rabbits [109].

Sanzgiri et al. compared various systems of methylprednisolone (MP); esters of MP with Gelrite eye drops, Gellan-MP film, and Gellan film with dispersed MP. Gellan eye drops provided better performance because they afforded the advantage of faster gelation over a high surface area in eye, whereas the results obtained with the Gellan-MP film seemed to indicate that the gelation at the surface of the film occurred very slowly, and the surface of release was not controlled [110].

Mourice and Srinivas reported use of gellan gum for permeation enhancement of the fluorescein in humans as compared to isotonic buffer solution. The ability of gelation at physiological $Ca^{2+}$ levels was used. The selected polymer significantly increased the duration of pilocarpine action to 10-h and carteolol to 8-h, reducing frequency of drug administration in case of carteolol [111].

Cohen et al. reported that sodium alginate aqueous solution could gel in the eye, without incorporation of external bivalent/polyvalent cations. The percentage of glucuronic acid residues in the polymer backbone estimate extent of alginate gelation and consequently the release of pilocarpine. Alginates having G content more than 65%, like Manugel DMB, instantaneously formed gels [112].

Lokhande et al. formulated pH-triggered *in-situ* gelling system of ciprofloxacin. Sol-gel transition of *in-situ* gelling system, in the cul-de-sac, results into the decreased pre-corneal elimination of drugs and thus improves bioavailability. Polymers used to prepare the formulation are Carbopol 934, HPMC-K4M, HPMC-E15V, and HPMC-E50LV. Parameters studied are clarity testing, drug content analysis, gelling characteristics, in-vitro release studies, sterility testing, ocular irritation studies, and FT-IR studies. No sign of ocular damage was observed during ocular irritation studies using albino rabbits [113].

Singh et al. formulated pH-sensitive hydrogels for ocular delivery to provide prolonged effect, increased bioavailability, and reduction in frequency of administration. Formulation is composed of drug timolol maleate, gelling agent poly(acrylic acid) (PAA), viscolizer HPMC and NaCl to maintain isotonicity. They also performed dug release studies using dynamic dialysis technique. Reduction in intraocular tension was studied using rabbit as experimental model [114].

Rathod et al. developed a controlled release *in-situ* gel of norfloxacin for ocular drug delivery. The polymers Carbopol 934 and HPMC used in formulation to enhance the contact time, to achieve controlled release, to decrease frequency of administration and to enhance therapeutic efficacy of drug. Parameters studied for *in-situ* gels are clarity testing, pH, drug content analysis, gelation, rheology, and sterility testing and drug release studies [115].

Shetty et al. developed naphazoline and antazoline *in-situ* gelling systems for ocular drug delivery and stability studies. The polymers used to prepare of *in-situ* gel are Carbopol 940 and HPMC K4M. The parameters studied for formulations are pH, isotonicity, gelling capacity, rheology, in-vitro release, sterility, and *in-vivo* studies. Formulation was found to be very stable at room temperature and at higher temperature [116].

Saxena et al. formulated the pH-sensitive hydrogels of levofloxacin hemihydrates for ophthalmic drug delivery. Levofloxacin is used for the treatment of acute conjunctivitis. β-cyclodextrin were added in addition to the polymers Carbopol 940 and HPMC, to increase the solubility of levo-floxacin. Ocular irritation studies in albino rabbits showed that the formulation does not cause damage to the cornea, iris, and conjunctiva [117].

Nayak et al. investigated the pH triggered *in-situ* ophthalmic gel of moxifloxacin hydrochloride. The moxifloxacin hydrochloride ophthalmic *in-situ* gel is prepared by the combination of polymers HPMC K15 and Carbopol 934. They performed various characterization methods such as FT-IR studies, clarity testing, drug content analysis, gelling characteristics, viscosity studies, in-vitro release studies, sterility testing, and ocular irritation study and accelerated stability studies. The formulation showed the good sustained release for the period of 8 hours [118].

Padma Preetha et al. designed *in-situ* gels of diclofenac sodium and studied various parameters such as pH, diffusion study, and sterility testing. They also studied effect in antimicrobial, antibacterial, and antifungal activities. The formulations studied showed the antimicrobial, antibacterial, and antifungal activity [119].

Patel et al. evaluated pH triggered *in-situ* ophthalmic gel formulation of ofloxacin. Ofloxacin is a fluoroquinolone. It is used most commonly to treat ocular infections like conjunctivitis, bacterial keratitis, and kerato-conjunctivitis. The ophthalmic gel was formulated by using the polymers Carbopol 974P, Noveon® AA-1 USP, and Polycarbophil (120).

Nagalakshmi et al. investigated stimuli sensitive pH triggered *in-situ* gelling system of fluconazole delivered by ocular route. Polymers used to prepare formulation are HPMC and Carbopol 940. Formulation studied showed good sustained release property as compared to other formulations for the period of 8 hours [121].

Kanoujia et al. studied application of a novel pH-triggered *in-situ* gelling ocular system containing gatifloxacin. The gatifloxacin is a fluoro-quinolone used to treat eye infections. Biodegradable polymers (HPMC, HPMC K15M, and Carbopol 940) were used [122].

## 3.6  APPLICATION OF BIO RESPONSIVE POLYMERS IN PARENTERAL DRUG DELIVERY

The parenteral administration route is one of the most efficient and is useful in delivering drugs with poor bioavailability. The development of new injectables has been essential to overcome the lacunae in existing drug delivery platforms [123]. Due to ease of investment, less complicated fabrication and easy manufacturing and delivery of sensitive materials, bioresponsive drug systems are future of novel administration comprising less investment and manufacturing cost [124]. A large number of *in-situ* forming and triggered release systems have been researched for parenteral drug delivery.

*In-situ* cross-linked systems are advantageous due to the possibility of controlling the diffusion of hydrophilic molecules. Such a system can release peptides and proteins for longer duration. These hydrogels are contraindicated to sites accessible to a light source, because they could form difficulties after injecting into body. Polymers containing double bonds and free radical-initiation are necessary in preparation of *in-situ* cross linked systems. Incorporation of benzoyl peroxide, which initiates free radical reaction, is necessary along with biodegradable polymers such as D, L-lactide, or L-lactide with E-caprolactone. However, benzoyl peroxide can induce tumor promotion [125].

*In-situ* polymer precipitation based systems are those where the drug is incorporated in solution of water insoluble, biodegradable polymer in biocompatible organic solvent. When such formulation is injected, the water miscible organic solvent dissipates and water penetrate to organic layer which results in phase separation and precipitation of polymer, at site of injection. Such a method was designed by ARTIX and designated as Atrigel technology [126]. Example of such technology is EligardTM, which contain the luteinizing hormone releasing hormone agonist leup-rolide acetate and poly(lactide-co-glycolic acid) (PLGA) dissolved in N-methyl-2-pyrrolidone (NMP) [126, 127]. This system is used to reduce testosterone levels in dogs. The problem with such system is the bursting out drug release after injecting drug. To control burst effect, four factors should be taken into consideration: concentration of polymers in solvents [128], molecular weight of polymers, solvent used [129] and addition of surfactants [130]. Brodbeck et al. studied that protein release is affected by type of solution formed. They studied NMP, triacetin, and ethyl benzoate ternary phase system with PLGA and water. The formulation with NMP Pyrrolidone shows rapid phase inversion and high burst while formula-tions containing triacetin and ethyl benzoate give low phase inversion and reduced burst of protein [131, 132].

Thermoresponsive polymers have been extensively researched in almost all routes of delivery. Such a drug delivery system is useful in delivery of water-insoluble drug substance such as paclitaxel for more than 50 days. Protein drugs can also be delivered as sustain release dosage form using thermally induced gel type delivery system. The stability of protein in aqueous solution, shelf life of formulation and bioavailability data for proteins are under investigation [133].

Thermoplastic or thermo softening pastes are semisolid polymers given in molten form and forms depot by solidifying at body temperature. The melting point of polymers ranges from 25–65°C and its viscosity ranges from 0.05–0.8 dl/g [134, 135]. Below the viscosity 0.05 dl/g, no delayed release can be observed and above viscosity 0.8 dl/g, the polymer cannot be injected through needle. Drugs can be incorporated into the polymer without using any solvent. Bioresponsive thermoplastic paste can be made up of D, L-lactide, glycolide, E-caprolactone, or orthoesters as polymers [134–136]. The applications of such monomers or co-polymers are, in preparation of sutures [137], ocular implants [138, 139] soft tissue repair [140], etc.

Zhang et al. designed a thermoplastic ABA triblock polymer system comprising poly(D,L-lactide)-poly(ethylene glycol)-poly(D,L-lactide) and blend of ABA triblock copolymer and polycaprolactone for efficient delivery of taxol within tumor resection sites. It was observed that both the systems ensured release of taxol for more than 60 days but at a very slow rate. However, this system required injection temperature at least 60°C which led to painful injections and necrosis at the site of injection [141].

Another example of thermosensitive 'Smart polymer' for controlled release of heparin is alginate-hydroxy propyl cellulose-based microbeads for which the release profiles were studied at different temperatures and various alginate/hydroxypropyl cellulose compositions. Such microbeads are spherical in shape with 3 micron diameter and contain encapsulated heparin obtained through emulsification method. The internal structure, surface structure, and morphology of microbeads were estimated by fluorescence microscopy, scanning electron microscopy and atomic force microscopy respectively. Lower critical solution temperature (LCST) of the systems was measured and release profile of heparin complex method showed three-stage sustained release for at least 16 days at 37°C. The release was correlated with size of pores present on surface of microbeads and can be controlled by temperature and composition of Alginate/ Hydroxy Propyl Cellulose microbeads [142].

Recent studies suggest that, poly[2-(2-ethoxy) ethoxyethyl vinyl ether (EOEOVE)] which is having LCST around 40°C, can be incorporated into liposomes and show good temperature-sensitive properties. Such type of thermosensitive polymer can be used in the preparation of tumor-specific thermosensitive liposome preparation containing DOX. For efficient delivery of DOX, liposomes consisting of PEG-lipid, egg yolk phosphatidylcholine, cholesterol, and copoly(EOEOVE-block-octadecyl vinyl ether) should be prepared. Such liposomes should be synthesized as poly(EOEOVE) comprising anchors for fixation on liposome membrane. These copolymer incorporated stable liposomes retain DOX at physiological temperature and efficiently release DOX above 40°C and release it within 1 min at 45°C. The copolymer modified liposomes loaded with DOX exhibit similarity in circulation and bio-distribution to PEG-modified liposomes, and can be injected intravenously into tumor-bearing mice. When the tumor site heated to 45°C for 10 min, tumor growth can be inhibited within 6–12 hours after injection, but produces slight tumor suppressive effect if mild heating is not applied to the target site. The temperature-sensitive properties of copolymer incorporated liposomes might contribute to the development of effective tumor-selective chemotherapy [143].

Suishaa et al. evaluated thermoreversible gels formed by xyloglucan polysaccharide derived from tamarind seed as a sustained-release vehicle for the administration of mitomycin C intraperitoneally. Administration of mitomycin C to rats in a 1.5% (w/w) xyloglucan gel intraperitoneally resulted in a broad concentration-time profile for this drug in ascites and the plasma over a 3-h time period as compared to a narrow peak and rapid disappearance from both sites after intraperitoneal administration of drug as a solution [144].

Ding et al. designed injectable hydrogels based on glycol CS and benzaldehyde-capped poly(ethylene glycol)-*block*-poly(propylene glycol)-*block*-poly(ethylene glycol) (PEO-PPO-PEO). *In vivo* tests using a rat model demonstrated that the hydrogel underwent a sol-gel transition at physiological conditions. These hydrogels have the ability to encapsulate both hydrophilic and hydrophobic drugs and can control the release profile by varying temperature or pH [145].

Salehi et al. developed an injectable hydrogel system of PNIPAm, acrylamide, and vinyl pyrrolidone to incorporate naltrexone, an opiate receptor antagonist. The swelling ratios of the gel increased with decrease in pH from 8.5 to 7.4 and decreased with increase in temperature from 25°C to 37°C. They also perform *in vitro* release studies where a low burst effect and a slow-release profile of naltrexone was noted for 28 days [146].

Stimuli-responsive materials have made it possible for drugs to be released in the acidic tumor and intracellular microenvironments, thus sparking increased application in chemotherapy.

Zhou et al. formulated the conjugates of a PEG polymer backbone with multiple DOX molecules linked to it. The conjugates have shown a significantly faster drug release at pH 6.0 than at pH 7.4. This is particularly useful for intracellular delivery. The cellular release of DOX was brought about by the cleavage of the hydrazone linkages. This allowed the drug to be accumulated in the nuclear compartment. The *in-vivo* studies on the conjugates confirmed a longer plasma half-life and tumor accumulation with intravenous administration [147].

The advances in the drug delivery systems have now made it more convenient and precise to deliver nucleic acids. Ghosn et al. investigated imidazole-modified CS for siRNA delivery. Post intravenous injection, the imidazole-modified CS siRNA complexes in mice, a marked knockdown of a target enzyme in both lung and liver was observed and that too at a lower dose [148].

Study on a reversibly-PEGylated diblock copolymer, poly(aspartate-hydrazide-poly(ethylene glycol))-block-poly(aspartate-diaminoethane) (p[Asp(Hyd-PEG)]-b-p[Asp(DET)]) was explored for enhanced gene transfection. The distinctive molecular architecture was because of the poly(aspartamide) backbone. The first block, p[Asp(Hyd)], offered sites for multi-PEG conjugations while the second block, p[Asp(DET)], allowed for DNA condensation and endosomal escape. The so formed polyplexes had a size below 100 nm shielded by a PEG layer. The polyplex demonstrated significantly higher transfection efficiency and pH sensitivity was seen at pH 5 [149].

Amphiphilic chondroitin sulfate-histamine conjugate were synthesized and formulated as nanoparticles in an aqueous medium with size below 150 nm and low critical micelle concentration, i.e., 0.05 mg/L. The nanoparticles displayed pH-responsive behavior upon reducing the pH of the microenvironment. A model drug, DOX was incorporated and the system exhibited a specific on-off switch drug release behavior, triggering drug release in intracellular endosomes while sealing it from neutral surroundings (blood circulation or extracellular matrix (ECM)). Advanced studies helped to ascertain its safety in cell-lines. The on-off modality offered the flexibility of delivering the drugs to tumor cells specifically [150].

Although intracellular delivery was thought as the best option for targeting chemotherapy, but it too has its own disadvantages as heterogeneity among the cancer cells may become a constraint.

Lee et al. developed a novel polymeric micelle as an alternative to cell-specific cancer-targeting strategies. The micelle was made up of two block copolymers of poly(L-lactic acid)-b-poly(ethylene glycol)-b-poly (L-histidine)-TAT (transactivator of transcription) and poly(L-histidine) -b-poly(ethylene glycol). The micelle which was formed by dialysis method had a size below 100 nm and had payload of around 15% DOX. The micelle surface protected the cell penetrating peptide in circulation and exposes it only at acidic tumor extracellular pH. This helps the micelle to be internalized. The micelle core disintegrated in endosomal pH to release the drug. The developed system ensured higher concentration of the drug in the cytosol. Cell-line studies in multidrug-resistant cell lines and *in-vivo* studies validated the platform technology [151].

Another similar approach was adopted by Wu et al. wherein they prepared tumor-targeting peptide (AP peptide; CRKRLDRN) conjugated pH-responsive polymeric micelles. The micelles were prepared from AP peptide conjugated PEG-poly(d,l-lactic acid) block copolymer (AP-PEG-PLA) within the pH-responsive micelles of methyl ether poly(ethylene glycol) (MPEG)-poly(beta-amino ester) (PAE) block copolymer (MPEG-PAE). They showed micellization and demicellization at tumor-specific pH. *In vivo* studies in MDA-MB231 human breast tumor-bearing mice with DOX loaded in the described micellar system displayed excellent targeting capability [152].

Hydrolytic cleavage was used in designing polymer backbones conjugated with anticancer drugs. The derivatized paclitaxel and DTX were attached to polymeric backbone via spacers and so formed conjugates were stable in physiological pH and released drug in pH-dependent fashion in endosomes. The polymer conjugates were devoid of toxicity and showed activity in mammary carcinoma and cell lymphoma models [153].

Pathological specificities like protein overexpression have been harnessed to improve cancer treatment efficacy and reduce toxicity of the stimuli sensitive polymeric micelle drug carriers. Folate moiety was conjugated in two block copolymers to utilize the folate receptor over-expression in malignancies. Folate-poly(ethylene glycol)-poly(aspartate-hydrazone-adriamycin) with gamma-carboxylic acid-activated folate and methoxy-poly(ethylene glycol)-poly(aspartate-hydrazone-adriamycin) without folate were taken to design the carriers. They did not affect accumulation in tumors but showed lower *in vivo* toxicity and higher antitumor activity over a broad range of the dosage [154].

Encapsulation of antigens in stimuli-responsive systems is of particular importance in vaccine delivery and immunotherapeutics. Cohen et al. explored a polyacrylamide hydrogel microparticles cross-linked with acid-labile moieties. The hydrophilic carriers which were of multiple size range could elicit an immune response equally. The microparticles were phagocytized by the antigen-presenting cells and antigen epitopes were presented to surmount a T cell response [155].

Shin et al. worked on the acid-catalyzed system, i.e., liposomes composed of polyethylene glycol (PEG) conjugated vinyl ether lipids along with, dioleoylphosphatidyl ethanolamine. Acid-catalyzed hydrolysis of the vinyl ether bond triggered the disassembly of these liposomes and removal of the sterically-stabilizing PEG layer. This prompted the drug release at pH < 5 and extending up to hours [156].

Bioreducible poly(amidoamine) with multiple disulfide linkages have been synthesized and investigated as nonviral gene vectors. The so formed polymers were stable in physiological pH but are rapidly degraded in reducing bioenvironment. The polymers are able to condense DNA with them and form positively charged nanocarriers sizing below 200 nm. They exhibit buffer capacities higher than polyethyleneimine which favors it for endosomal escape. In-vitro transfection efficiencies of the synthesized polymers were much higher than polyethyleneimine. They represent a good option for nontoxic, safe, and efficient nanocarriers for transfection of nucleic acids [157].

The matrix metalloproteinases (MMPs) are proteases which are over-expressed in tumor tissues. MMPs degrade biocompatible and non-immunogenic substrate gelatine, so gelatin coated mesoporous silica nanoparticles (MSPs) were designed for endogenous tumor microenvironment (TME)-triggered release of anticancer drugs with diminished systemic toxicity. The gelatin corona gives dual benefit of providing protective layer for preventing drug leakage and MMPs-digested substrates which are responsive in solid TME. DOX, a novel anticancer drug was loaded into mesopores of MMPs-degradable gelatin-coated MSPs which controllably release DOX selectively in cancerous cells while releasing the very minute amount of drug in normal cells due to less endocytosis. Such selectivity decreases systemic toxicity and side effect which is a drawback of cancer chemotherapy treatment. *In-vivo* studies suggested that the tumor growth of xenografted mice was significantly delayed without loss of body weight compared to free DOX. Hence, we can conclude that gelatin-MSPs as MMPs-degradable drug delivery system is one of a promising and less toxic delivery system *in vivo* [158].

The magnetic nanoparticles are used to enhance therapeutic activity and thermal stability of 1,3-bis (2-chloroethyl)-1-nitrosourea (BCNU), a compound used to treat brain tumors. Such drug carriers are prepared using polymer poly [aniline-co-N-(1-butyric acid) aniline] (SPAnH) coated on $Fe_3O_4$ cores. The nanoparticles have diameter of 89.2 ± 8.5 nm and show superparamagnetic properties. 1 mg of magnetic nanoparicle (bound-BCNU-3) can be combined with maximum effective dose of 379.34 µg BCNU and was more stable than free-BCNU when stored at 4°C, 25°C, or 37°C. Bound-BCNU-3 can be used at targeted sites in-vitro and *in-vivo* using externally applied magnet and when applied to tumors, magnetic targeting increases retention and concentration of bound-BCNU-3. Such kind of drug delivery system gives promising result and effective tumor treatment using lower therapeutic dose and hence reduces side effects of chemotherapy [159].

One of futuristic therapeutic approach is nanotherapeutic strategy which consists of nanoparticulate drug carriers incorporated with a moiety that targets certain diseased cells. In recent studies, such a nanosystem is modified to accurately deliver anticancer agent camptothecin to cancer cells overexpressing epithelial growth factor receptor (EGFR). The parameters analyzed were the endocytosis of nanocarriers by cancer cells, the pathway of cellular uptake and subsequent intracellular controlled drug delivery. The modified nanocarriers showed comparatively higher camptothecin-drug load efficiency and higher cellular uptake through clathrin-mediated endocytosis by EGFR overexpressive cancer cells. The external magnetic stimulus mediated intracellular release of camptothecin was proved to be much promising and consists of higher therapeutic efficacy than free camptothecin drug. This study brings focus on multiple functions of nanotherapeutic treatment like cell-specific targeting, controlled cellular endocytosis, and magnetic responsive drug release [160].

The next example is of low-frequency ultrasound (LFUS), which releases drugs from nano sterically stabilized liposomes *in vitro*, without affecting the biological potency of drug. In this study, mice bearing well developed J6456 murine lymphoma tumors were injected nano sterically stabilized liposomes loaded with anticancer cisplatin intraperitoneally.

Then the drug release was studied by applying LFUS externally to the abdominal wall for 120 seconds. Approximately 70% of liposomal cisplatin was released in LFUS exposed tumors, while < 3% was released in tumor areas not exposed to LFUS. The BALB/c mice with C26 colon adenocarcinoma tumors were tested to study the therapeutic efficacy of LFUS induced localized drug release. After 24 hours of intravenous injection with nano sterically stabilized liposomes cisplatin in mice, the tumor was exposed to LFUS. The best-proven treatment was observed in the group treated with liposomal cisplatin combined with LFUS, compared to other groups such as free cisplatin with or without LFUS, or liposomal cisplatin without LFUS, or LFUS alone or no treatment. The tumor was observed stopped proliferating and then regressed gradually. Such type of work presents clinical applications of LFUS induced liposomal drug release [161].

[2,9,17,23-tetrakis-(1,6-hexanedithiol)phthalocyaninato]zinc (II) is found to be useful as photodynamic therapy agent. It belongs to second generation of photosensitizers acting against cellular photo-damage of cancer cells. Their suspected mechanism is production of reactive oxygen species and phototoxicity of the photosensitizer. The effect of phthalocyanine or phthalocyanine bound to gold nanoparticles or encapsulated in liposomes was studied on healthy fibroblast cells and breast cancer (MCF-7) cells. Studies suggested optimum phototoxic effect on cancer cells to be 4.5 Jcm (-2). The liposomes bounded phthalocyanine showed photodynamic effect and extensively damaged breast cancer cells, whereas gold nanoparticles slightly improved photodynamic therapy effect [162].

Though radiosensitizing effect of caffeine and other methylxanthine derivatives on cancer cells is promising, their clinical use is limited due to their toxicity, requirement of higher dose, poor solubility. This study consists of an evaluation of efficacy of a caffeine metabolite, 1-methylxanthine as a radiosensitizer and *in vivo* study of temperature sensitive liposomal 1-methylxanthine along with ionizing radiation and regional hyperthermia. Treatment of 1-methylxanthine sensitizes human colorectal and lung cancer cells to ionizing radiation. Temperature-sensitive liposomal 1-methylxanthine consisting DPPC:DMPC:DSPC (4:1:1 molar ratio) was developed to evaluate its capability *in vivo* to radiosensitize tumors which overcomes lethal toxicity of 1-methylxanthine and control drug release. The liposomes are of approximately 200 nm in diameter. The release of 1-methylxanthine from liposomes was temperature responsive.

The temperature-sensitive liposomal 1-methylxanthine was administered in xenograft tumor-bearing mice via i.p. route and was showing a delay of tumor growth. The temperature-sensitive liposomal 1-methylxanthine in combination with radiation and regional hyperthermia significantly reduces growth of tumor, which suggests that 1-methylxanthine is effectively used in suppression of radiation mediated tumor growth. Hence, it is concluded that temperature sensitive liposomal 1-methylxanthine is promising anticancer drug *in vivo* and also increases radiotherapeutic effectiveness and feasibility for clinical use [163].

pH-responsive release of anti-HIV microbicides in the presence of human semen fluid stimulant was studied by developing tenofovir or tenofovir disoproxil fumarate loaded nanoparticles which consists combination of noncytotoxic copolymers PLGA and methacrylic acid (Eudragit® S-100, or S-100). After preparation of nanoparticles different techniques like dynamic light scattering, spectrophotometry, transmission electron microscopy, and cellular viability assay/transepithelial electrical resistance are used for determination of characteristics like emulsification diffusion, size of nanoparticles, encapsulation efficiency, drug release profile, morphology, and cytotoxicity respectively. Fluorescence spectroscopy and confocal microscopy was used to elucidate cellular uptake. The nanoparticles have an average size of 250 nm, entrapment efficiency of 16.1% and 37.2% for tenofovir and tenofovir disoproxil fumarate respectively. With such system 4-fold increase in the drug release rate in presence of semen, fluid stimulant over 72 hours was observed. The PLGA/s-100 NPs are noncytotoxic for 48 hours to vaginal endocervical/epithelial cells and *Lactobacillus crispatus* at concentration of 10 mg/ml. The caveolin pathway mediated particle uptake (50% in 24 hours) by these vaginal cell lines, suggested promising use of PLGA/s-100 NPs as an alternative drug delivery system consisting intravaginal delivery of an anti-HIV/AIDS microbicide [164].

One of novel example of pH and temperature-sensitive hydrogel is (PAE-PCL-PEG-PCL-PAE) pentablock copolymer, which consist of poly(beta-amino ester)-poly(epsilon-caprolactone)-poly(ethylene glycol)-poly(epsilon-caprolactone)-poly(beta-amino ester) and was evaluated as a sustained insulin release injectable. Insulin loading into matrix can be done by ionically linking of insulin-PAE complex. *In vivo* release of insulin into male sprague-dawley rats can be studied by subcutaneously injecting different concentrations of insulin and copolymer. The insulin level was maintained constant for 15 days and further controlled by the amount of insulin loaded into copolymer and copolymer concentration in the hydrogel. Streptozotocin diabetic rat model was further investigated to study the effect of insulin-gel complex. After subcutaneous injection of insulin complex mixture into streptozotocin-induced diabetic rates, blood glucose levels (BGLs) and plasma insulin levels were analyzed. The results suggested that single injection of the complex mixture containing 10 mg/ml insulin in 30% copolymer solution can treat such diabetic rats for more than 1 week. Hence, it is concluded that this pH/temperature sensitive hydrogel complex system is promising and have good therapeutic potential [165].

Platinum (II) chemotherapy drugs are often prescribed but have disadvantages like poor pharmacokinetics, side effects and develop drug resistance easily. By developing amphiphilic, block copolymers based micelles; novel macromolecular platform for carrier-based delivery of such compounds can be developed. Condensation of the poly(ethylene oxide)-b-polymethacrylate anions by metal ions into core-shell of complex micelles and chemical cross-linking of polyion chains in the micelle cores can form soft polymeric nanocarriers, which can efficiently incorporate cisplatin with high loading capacity up to 42% w/w. Due to cross-linking micelles are stabilized against structural disintegration and premature drug release. For reversible entrapment of cisplatin, the micelle core should have carboxylate groups in its structure. Ultimately, the drug releases in a pH-responsive manner, without losing its biological activity. Such kind of stable cross-linked polymer micelles can effectively distribute platinum (II) with improved therapeutic potential [166].

## 3.7 CONCLUSION

Scientific literature is full of smart polymer-based drug delivery, diagnostic, and tissue engineering applications. However, not many platform technologies could make it to market. The reasons for such a large number of products which could not be translated lies in their toxicities. The safety considerations become all the more important because of the increasing trends towards intracellular targeting of biomolecular drugs. It has been observed that high molecular weight (*HMW*) polymers have been more successful in meeting the drug delivery challenges, but they are a burden on kidney and not excreted readily. It is presumed that diagnostic applications are likely to be more successful. But with a deeper and wider understanding of biological processes, it would be possible to pinpoint the targets for bioresponsive systems. Further, the development of newer and safer biomaterials may drive the future of the stimuli-responsive systems.

## KEYWORDS

- **age-related macular degeneration**
- **brain-derived neurotrophic factor**
- **chitosan**
- **ethylene glycol dimethacrylate**
- **glucagon-like peptide-1**
- **human neutrophil elastase**

## REFERENCES

1. Hoffman, A. S., (2013). Stimuli-responsive polymers: Biomedical applications and challenges for clinical translation. *Adv. Drug Delivery Rev., 65,* 11–16.
2. Almeida, H., Amaral, M. H., & Lobão, P., (2012). Temperature and pH stimuli-responsive polymers and their applications in controlled and self-regulated drug delivery. *J. Appl. Pharm. Sci., 2,* 1–10.
3. Liu, D., Yang, F., Xiong, F., & Gu, N., (2016). The smart drug delivery system and its clinical potential. *Theranostics, 6,* 1301–1323.
4. Suzuki, A., & Tanaka, T., (1990). Phase transition in polymer gels induced by visible light. *Nature, 346,* 341–347.
5. Kwon, S. S., Kong, B. J., & Park, S. N., (2015). Physicochemical properties of pH-sensitive hydrogels based on hydroxyethyl cellulose-hyaluronic acid and for applications as transdermal delivery systems for skin lesions. *Eur. J. Pharm. Biopharm., 92,* 146–154.
6. Aimetti, A. A., Tibbitt, M. W., & Anseth, K. S., (2009). Human neutrophil elastase responsive delivery from poly(ethylene glycol) hydrogels. *Biomacromolecules, 10,* 1481–1489.
7. Wang, C., Ye, Y., Hochu, G. M., Sadeghifar, H., & Gu, Z., (2016). Enhanced cancer immunotherapy by microneedle patch-assisted delivery of anti-PD1 antibody. *Nano Lett., 16,* 2331–2340.
8. Ye, Y., Wang, J., Hu, Q., Hochu, G. M., Xin, H., Wang, C., & Gu, Z., (2016). Synergistic transcutaneous immunotherapy enhances antitumor immune responses through delivery of checkpoint inhibitors. *ACS Nano, 10,* 8951–8963.

9. Indermun, S., Choonara, Y. E., Pradeep, K., Toit, D. L. C., Modi, G., Luttge, R., & Pillay, V., (2013). An interfacially plasticized electro-responsive hydrogel for transdermal electro-activated and modulated (TEAM) drug delivery. *Int. J. Pharm., 462,* 51–65.

10. Indulekha, S., Arunkumar, P., Bahadur, D., & Srivastava, R., (2016). Thermoresponsive polymeric gel as an on-demand transdermal drug delivery system for pain management. *Mater Sci. Eng. C. Mater Biol. Appl., 62,* 111–122.

11. Escobar-Chávez, J. J., López-Cervantes, M., Naïk, A., Kalia, Y. N., Quintanar-Guerrero, A., & Ganem-Quintanar, A., (2006). Applications of thermo-reversible pluronic F-127 gels in pharmaceutical formulations. *J. Pharm. Pharm. Sci., 9,* 331–358.

12. Choi, J. S., Kim, H. S., & Yoo, H. S., (2015). Electrospinning strategies of drug-incorporated nanofibrous mats for wound recovery. *Drug Delivery Transl. Res., 5,* 131–145.

13. You, J. O., Rafat, M., Almeda, D., Maldonado, N., Guo, P., Nabzdyk, C. S., et al., (2015). pH-responsive scaffolds generate a pro-healing response. *Biomaterials, 57,* 21–32.

14. Okino, H., Nakayama, Y., Tanaka, M., & Matsuda, T., (2002). *In situ* hydrogelation of photocurable gelatin and drug release. *J. Biomed. Mater. Res., 59,* 231–245.

15. Yu, J., Zhang, Y., Kahkoska, A. R., & Gu, Z., (2017). Bioresponsive transcutaneous patches. *Curr. Opin. Biotechnol., 48,* 21–32.

16. Donnelly, R. F., Singh, T. R. R., Morrow, D. I., & Woolfson, A. D., (2012). *Microneedle-Mediated Transdermal and Intradermal Drug Delivery.* John Wiley & Sons: Chichester, UK.

17. Hardy, J. G., Larraneta, E., Donnelly, R. F., McGoldrick, N., Migalska, K., Mc Crudden, M. T., Irwin, N. J., Donnelly, L., & McCoy, C. P., (2016). Hydrogel-forming microneedle arrays made from light-responsive materials for on-demand transdermal drug delivery. *Mol. Pharmaceutics, 13,* 901–914.

18. Ke, C. J., Lin, Y. J., Hu, Y. C., Chiang, W. L., Chen, K. J., Yang, W. C., Liu, H. L., Fu, C. C., & Sung, H. W., (2012). Multidrug release based on microneedle arrays filled with pH-responsive PLGA hollow microspheres. *Biomaterials, 33,* 5156–5165.

19. Van Der Maaden, K., Yu, H., Sliedregt, K., Zwier, R., Leboux, R., Oguri, M., Kros, A., Jiskoot, W., & Bouwstra, J. A., (2013). Nanolayered chemical modification of silicon surfaces with ionizable surface groups for pH-triggered protein adsorption and release: Application to microneedles. *J. Mater. Chem. B., 1,* 4461–4477.

20. Yu, J., Zhang, Y., Ye, Y., DiSanto, R., Sun, W., Ranson, D., Ligler, F. S., Buse, J. B., & Gu, Z., (2015). Microneedle-array patches loaded with hypoxia-sensitive vesicles provide fast glucose-responsive insulin delivery. *Proc. Natl. Acad. Sci. U.S.A., 112,* 8260–8265.

21. Ye, Y., Yu, J., Wang, C., Nguyen, N. Y., Walker, G. M., Buse, J. B., & Gu, Z., (2016). Microneedles integrated with pancreatic cells and synthetic glucose-signal amplifiers for smart insulin delivery. *Adv. Mater., 28,* 3115–3121.

22. Hu, X., Yu, J., Qian, C., Lu, Y., Kahkoska, A. R., Xie, Z., Jing, X., Buse, J. B., & Gu, Z., (2017). $H_2O_2$-responsive vesicles integrated with transcutaneous patches for glucose-mediated insulin delivery. *ACS Nano, 11,* 611–620.

23. Yu, J., Qian, C., Zhang, Y., Cui, Z., Zhu, Y., Shen, Q., Ligler, F. S., Buse, J. B., & Gu, Z., (2017). Hypoxia and $H_2O_2$ dual-sensitive vesicles for enhanced glucose-responsive insulin delivery. *Nano Lett., 17,* 731–739.

24. Zhang, Y., Yu, J., Wang, J., Hanne, N. J., Cui, Z., Qian, C., Wang, C., Xin, H., Cole, J. H., & Gallippi, C. M., (2017). Thrombin-responsive transcutaneous patch for auto-anticoagulant regulation. *Adv. Mater.,* 29. doi: 10.1002/adma.201604043.

25. Kim, N. W., Lee, M. S., Kim, K. R., Lee, J. E., Lee, K., Park, J. S., et al., (2014). Polyplex-releasing microneedles for enhanced cutaneous delivery of DNA vaccine. *J. Controlled Release, 179*, 11–17.

26. Samani, S. M., Zolfaghari, N., Parhizkar, E., & Ahmadi, F., (2017). Orally fast disintegrating tablets of Sumatriptan loaded Eudragit E microparticles: A promising choice for taste masking. *Int. J. Pharm. Sci. Res., 8*, 2003–2012.

27. Yan, Y. D., Woo, J. S., Kang, J. H., Yong, C. S., & Chui, H. G., (2010). Preparation and evaluation of tastemasked donepezil hydrochloride orally disintegrating tablets. *Biol. Pharm. Bull., 33*, 1361–1370.

28. Subudhi, M. B., Jain, A., Hurkat, P., Shilpi, S., Gulbake, A., & Jain, S. K., (2015). Eudragit S100 coated citrus pectin nanoparticles for colon targeting of 5-fluorouracil. *Materials, 8*, 831–849.

29. Zhang, S., Ermann, J., Succi, M. D., Zhou, A., Hamilton, M. J., Cao, B., et al., (2015). An inflammation-targeting hydrogel for local drug delivery in inflammatory bowel disease. *Sci. Transl. Med., 7*(300), 1–10.

30. Koetting, M. C., Guido, J. F., Gupta, M., Zhang, A., & Peppas, N. A., (2015). pH-responsive and enzymatically-responsive hydrogel microparticles for the oral delivery of therapeutic proteins: Effects of protein size, crosslinking density, and hydrogel degradation on protein delivery. *J. Controlled Release, 221*, 11–25.

31. Koetting, M. C., & Peppas, N. A., (2014). pH-Responsive poly(itaconic acid-co-N-vinylpyrrolidone) hydrogels with reduced ionic strength loading solutions offer improved oral delivery potential for high isoelectric point-exhibiting therapeutic proteins. *Int. J. Pharm., 471*, 81–91.

32. Foss, A. C., Goto, T., Morishita, M., & Peppas, N. A., (2004). Development of acrylic-based copolymers for oral insulin delivery. *Eur. J. Pharm. Biopharm., 57*, 161–169.

33. Sung, H. W., Sonaje, K., Liao, Z. X., Hsu, L. W., & Chuang, E. Y., (2012). pH-responsive nanoparticles shelled with chitosan for oral delivery of insulin: From mechanism to therapeutic applications. *Acc. Chem. Res., 45*, 611–629.

34. Wilson, D. S., Dalmasso, G., Wang, L., Sitaraman, S. V., Merlin, D., & Murthy, N., (2010). Orally delivered thioketal nanoparticles loaded with TNF-α-siRNA target inflammation and inhibit gene expression in the intestines. *Nat. Mater, 9*, 923–928.

35. Haseeb, M.T., Hussain, M.A., Bashir, S., Ashraf, M.U., Ahmad, N. (2017). Evaluation of superabsorbent linseed-polysaccharides as a novel stimuli-responsive oral sustained release drug delivery system. *Drug Dev. Ind. Pharm., 43*, 401–420.

36. Dai, J. D., Nagai, T., Wang, X. Q., Zhang, T., Meng, M., & Zhang, Q., (2004). pH-sensitive nanoparticles for improving the oral bioavailability of cyclosporine A. *Int. J. Pharm., 280*, 221–240.

37. Satturwar, P., Eddine, M. N., Ravenelle, F., & Leroux, J. C., (2007). pH-responsive polymeric micelles of poly(ethylene glycol)-b-poly(alkyl(meth)acrylate-co-methacrylic acid), influence of the copolymer composition on self-assembling properties and release of Candesartan cilexetil. *Eur. J. Pharm. Biopharm., 65*, 371–387.

38. Osada, K., Christie, R. J., & Kataoka, K., (2009). Polymeric micelles from poly (ethylene glycol)-poly(amino acid) block copolymer for drug and gene delivery. *J. R. Soc., Interface, 6,* S325–S339.

39. An, J., Geib, S. J., & Rosi, N. L., (2009). Cation-triggered drug release from a porous zinc-adeninate metal-organic framework. *Journal of American Chemical Society, 131,* 8371–8377.

40. Cohen, S., Coué, G., Beno, D., Korenstein, R., & Engbersen, J. F., (2012). Bioreducible poly(amidoamine)s as carriers for intracellular protein delivery to intestinal cells. *Biomaterials, 33,* 611–623.

41. Qin, S., Geng, Y., Discher, D. E., & Yang, (2006). Temperature-controlled assembly and release from polymer vesicles of poly(ethylene oxide)-*block*- poly(*N*-isopropylacrylamide). *Adv. Mater, 18,* 2901–2909.

42. Qu, W., Li, Y., Hovgaard, L., Li, S., Dai, W., Wang, J., Zhang, X., & Zhang, Q., (2012). A silica-based pH-sensitive nanomatrix system improves the oral absorption and efficacy of incretin hormone glucagon-like peptide-1. *Int. J. Nanomed., 7,* 4981–4994.

43. Lee, C. H., Lo, L. W., Mou, C. Y., & Yang, C. S., (2008). Synthesis and characterization of positive-charge functionalized mesoporous silica nanoparticles for oral drug delivery of an anti-inflammatory drug. *Adv. Funct. Mater., 18,* 3281–3292.

44. Collin, F., Karkare, S., & Maxwell, A., (2011). Exploiting bacterial DNA gyrase as a drug target: current state and perspectives. *Appl. Microbiol. Biotechnol., 92,* 471–497.

45. Eeckman, F., Moes, A. J., & Amighi, K., (2004). Synthesis and characterization of thermosensitive copolymers for oral controlled drug delivery. *Eur. Polym. J., 40,* 873–881.

46. Gröning, R., Berntgen, M., & Georgarakis, M., (1998). Acyclovir serum concentrations following peroral administration of magnetic depot tablets and the influence of extracorporal magnets to control gastrointestinal transit. *Eur. J. Pharm. Biopharm., 46,* 281–291.

47. Wang, C. Y., Yang, C. H., Lin, Y. S., Chen, C. H., & Huang, K. S., (2012). Anti-inflammatory effect with high intensity focused ultrasound-mediated pulsatile delivery of diclofenac. *Biomaterials, 33,* 1541–1553.

48. Mura, S., Nicolas, J., & Couvreur, P., (2013). Stimuli-responsive nanocarriers for drug delivery. *Nat. Mater, 12,* 991–1003.

49. Ganta, Y. S., Devalapally, H., Shahiwala, A., & Amiji, M., (2008). A review of stimuli-responsive nanocarriers for drug and gene delivery. *J. Controlled Release, 126,* 181–204.

50. Karavasili, C., & Fatouros, D. G., (2016). Smart materials: *In situ* gel-forming systems for nasal delivery. *Drug Discovery Today, 21*(1), 151–166.

51. Chonkar, A., Nayak, U., & Udupa, N., (2015). Smart polymers in nasal drug delivery. *Indian J. Pharm. Sci., 77*(4), 361–375.

52. Majithiya, R. J., Ghosh, P. K., Umrethia, M. L., & Murthy, R. S. R., (2006). Thermoreversible-mucoadhesive gel for nasal delivery of Sumatriptan. *AAPS Pharm. Sci. Tech., 7,* E80–E86.

53. Zaki, N. M., Awad, G. A., Mortada, N. D., & Abd Elhady, S. S., (2007). Enhanced bioavailability of metoclopramide HCl by intranasal administration of a mucoadhesive *in situ* gel with modulated rheological and mucociliary transport properties. *Eur. J. Pharm. Sci., 32,* 291–307.

54. Li, C., Li, C., Liu, Z., Li, Q., Yan, X., Liu, Y., & Lu, W., (2014). Enhancement in bioavailability of ketorolac tromethamine via intranasal *in situ* hydrogel based on poloxamer 407 and carrageenan. *Int. J. Pharm., 474*, 123–133.

55. Kempwade, A., & Taranalli, A., (2014). Formulation and evaluation of thermoreversible, mucoadhesive *in situ* intranasal gel of rizatriptan benzoate. *J. Sol-Gel Sci. Technol., 72*, 41–48.

56. Qian, S., Wong, Y. C., & Zuo, Z., (2014). Development, characterization, and application of *in situ* gel systems for intranasal delivery of tacrine. *International Journal of Pharmaceutics, 468*, 271–282.

57. Gabal, Y. M., Kamel, A. O., Sammour, O. A., & Elshafeey, A. H., (2014). Effect of surface charge on the brain delivery of nanostructured lipid carrier's *in situ* gels via the nasal route. *Int. J. Pharm., 473*, 442–457.

58. Perez, A. P., Weilenmann, C. M., Romero, E. L., & Morilla, M. J., (2012). Increased brain radioactivity by intranasal [32]P-labeled siRNA dendriplexes within *in situ*-forming mucoadhesive gels. *Int. J. Nanomed., 7*, 1371–1385.

59. Jose, S., Ansa, C. R., Cinu, T. A., Chacko, A. J., Aleykutty, N. A., Ferreira, S. V., & Souto, E. B., (2013). Thermo-sensitive gels containing lorazepam microspheres for intranasal brain targeting. *Int. J. Pharm., 441*, 516- 526.

60. Bhandwalkar, M. J., & Avachat, A. M., (2013). Thermoreversible nasal *in situ* gel of venlafaxine hydrochloride: formulation, characterization, and pharmacodynamic evaluation. *AAPS Pharm. Sci. Tech., 14*, 101–110.

61. Singh, R. M. P., Kumar, A., & Pathak, K., (2013). Thermally triggered mucoadhesive *in situ* gel of loratadine: β-cyclodextrin complex for nasal delivery. *AAPS Pharm. Sci. Tech., 14*, 412–424.

62. Basu, S., & Bandyopadhyay A. K., (2010). Development and characterization of mucoadhesive *in situ* nasal gel of midazolam prepared with *Ficus carica* mucilage. *AAPS Pharm. Sci. Tech., 11*, 1223–1231.

63. Mahajan, H. S., Tyagi, V., Lohiya, G., & Nerkar, P., (2012). Thermally reversible xyloglucan gels as vehicles for nasal drug delivery. *Drug Delivery, 19*, 271–276.

64. Bernkop-Schnurch, A., & Dunnhaupt, S., (2012). Chitosan-based drug delivery systems. *Eur. J. Pharm. Biopharm., 81*, 463–469.

65. Cho, H. J., Balakrishnan, P., Park, E. K., Song, K. W., Hong, S. S., Jang, T. Y., et al., (2011). Poloxamer/cyclodextrin/chitosan-based thermoreversible gel for intranasal delivery of fexofenadine hydrochloride. *J. Pharm. Sci., 100*, 681–691.

66. Nazar, H., Fatouros D. G., Van Der Merwe, S. M., Bouropoulos, N., Avgouropoulos, G., Tsiouklis, J., & Roldo, M., (2011). Thermosensitive hydrogels for nasal drug delivery: The formulation and characterization of systems based on N-trimethyl chitosan chloride. *Eur. J. Pharm. Biopharm., 77*, 221–232.

67. Nazar, H., Caliceti, P., Carpenter, B., El-Mallah, A. I., Fatouros, D. G., Roldo, M., Van Der Merwe, S. M., & Tsiouklis, J., (2013). A once-a-day dosage form for the delivery of insulin through the nasal route: *In vitro* assessment and *in vivo* evaluation. *Biomater. Sci., 1*, 301–314.

68. Wu, J., Wei, W., Wang, L. Y., Su, Z. G., & Ma, G. H., (2007). A thermosensitive hydrogel based on quaternized chitosan and poly(ethylene glycol) for nasal drug delivery system. *Biomaterials, 28*, 2220–2232.

69. Grasdalen, H., & Smidsroed, O., (1987). Gelation of gellan gum. *Carbohydr. Polym.*, *7*, 371–393.

70. Cao, S. L., Zhang, Q. Z., & Jiang, X. G., (2007). Preparation of ion-activated *in situ* gel systems of scopolamine hydrobromide and evaluation of its antimotion sickness efficacy. *Acta Pharmacol. Sin.*, *28*, 581–590.

71. Cao, S. L., Ren, X. W., Zhang, Q. Z., Chen, E., Xu, F., Chen, J., Liu, L. C., & Jiang, X. G., (2009). *In situ* gel based on gellan gum as new carrier for nasal administration of mometasone furoate. *Int. J. Pharm.*, *365*, 101–115.

72. Galgatte, U. C., Kumbhar, A. B., & Chaudhari, P. D., (2014). Development of *in situ* gel for nasal delivery: Design, optimization, *in vitro* and *in vivo* evaluation. *Drug Delivery*, *21*, 61–73.

73. Wang, S., Chen, P., Zhang, L., Yang, C., & Zhai, G., (2012). Formulation and evaluation of microemulsion-based *in situ* ion-sensitive gelling systems for intranasal administration of curcumin. *J. Drug Targeting*, *20*, 831–840.

74. Saindane, N. S., Pagar, K. P., & Vavia, P. R., (2013). Nanosuspension based *in situ* gelling nasal spray of carvedilol: Development, *in vitro* and *in vivo* characterization. *AAPS Pharm. Sci. Tech.*, *14*, 181–199.

75. Castile, J., Cheng, Y. H., Simmons, B., Perelman, M., Smith, A., & Watts, P., (2013). Development of *in vitro* models to demonstrate the ability of PecSys®, an *in situ* nasal gelling technology, to reduce nasal run-off and drip. *Drug Dev. Ind. Pharm.*, *39*, 811–824.

76. McConaughy, S. D., Stroud, P. A., Boudreaux, P., Hesterb, R. D., & McCormick, C. L., (2008). Structural characterization and solution properties of a galacturonate polysaccharide derived from Aloe Vera capable of *in situ* gelation. *Biomacromolecules*, *9*, 471–480.

77. McConaughy, S. D., Kirkland, S. E., Treat, N. J., Stroud, P. A., & McCormick, C. L., (2008). Tailoring the network properties of Ca2+ crosslinked Aloe Vera polysaccharide hydrogels for *in situ* release of therapeutic agents. *Biomacromolecules*, *9*, 3277.

78. Bertram, U., & Bodmeier, R., (2012). Effect of polymer molecular weight and of polymer blends on the properties of rapidly gelling nasal inserts. *Drug Dev. Ind. Pharm.*, *38*, 651–669.

79. Bertram, U., & Bodmeier, R., (2006). *In situ* gelling, bioadhesive nasal inserts for extended drug delivery: *In vitro* characterization of a new nasal dosage form. *Eur. J. Pharm. Sci.*, *27*, 61–71.

80. Bertram, U., Bernard, M.C., Haensler, J., Maincent, P., Bodmeier, R. (2010). *In situ* gelling nasal inserts for influenza vaccine delivery. *Drug Dev. Ind. Pharm.*, *2010*, *36*, 581–593.

81. Farid, R. M., Etman, M. A., Nada, A. H., Azeem, A. E., & Ebian, R., (2013). Formulation and *in vitro* evaluation of salbutamol sulphate *in situ* gelling nasal inserts. *AAPS Pharm. Sci. Tech.*, *14*, 711–718.

82. Nakamura, K., Maitani, Y., Lowman, A. M., Takayama, K., Peppas, N. A., & Nagai, T., (1999). Uptake and release of budesonide from mucoadhesive, pH-sensitive copolymers, and their application to nasal delivery. *J. Controlled Release*, *61*, 321–335.

83. Aikawa, K., Mitsutake, A., Uda, H., Tanaka, S., Shimamura, H., Aramaki, Y., & Tsuchiyab, S., (1998). Drug release from pH-response polyvinylacetal diethylaminoacetate hydrogel, and application to nasal delivery. *Int. J. Pharm.*, *168*, 181–188.

84. Ghosn, B., Singh, A., Li, M., Vlassov, A. V., Burnett, C., Puri, N., & Roy, K., (2010). Efficient gene silencing in lungs and liver using imidazole-modified chitosan as a nanocarrier for small interfering RNA. *Oligonucleotides, 20,* 161–172.

85. Li, Z., Xiong, F., He, J., & Wang, J., (2016). Surface-functionalized, pH-responsive poly(lactic-co-glycolic acid)-based microparticles for intranasal vaccine delivery: Effect of surface modification with chitosan and mannan. *Eur. J. Pharm. Biopharm., 109,* 21–34.

86. Su, X., Fricke, J., Kavanagh, D. J., & Irvine, D. J., (2011). *In vitro* and *in vivo* mRNA delivery using lipid-enveloped pH-responsive polymer nanoparticles. *Mol. Pharmaceutics, 8,* 771–787.

87. Xi, J., Zhang, Z., Si, X. A., Yang, J., & Deng, W., (2016). Optimization of magnetophoretic-guided drug delivery to the olfactory region in a human nose model. *Biomech. Model. Mechanobiol., 15,* 871–891.

88. Xi, J., Zhang, Z., & Si, X. A., (2015). Improving intranasal delivery of neurological nanomedicine to the olfactory region using magnetophoretic guidance of microsphere carriers. *Int. J. Nanomed., 10,* 1211–1222.

89. Chen, H., Zong, G., Yang, X., Getachew, H., Acosta, C., Sanchez, C. S., & Konofagou, E. E., (2016). *Sci. Rep., 6,* 28599.

90. Yasin, M. N., Svirskis, D., Seyfoddin, A., & Rupenthal, I. D., (2014). Implants for drug delivery to the posterior segment of the eye: A focus on stimuli-responsive and tunable release systems. *J. Controlled Release, 28,* 201–221.

91. Mahlumba, P., Choonara, Y. E., Kumar, P., Toit, L. C., & Pillay, V., (2016). Stimuli-responsive polymeric systems for controlled protein and peptide delivery: Future implications for ocular delivery. *Molecules, 21,* 1002, 1–21.

92. Jiang, J. Q., Tong, X., Morris, D., & Zhao, Y., (2006). Toward photocontrolled release using light-dissociable block copolymer micelles. *Macromolecules, 39,* 4631–4640.

93. Juzenas, P., Juzeniene, A., Kaalhus, O., Iani, V., & Moan, J., (2002). Noninvasive fluorescence excitation spectroscopy during application of 5-aminolevulinic acid *in vivo*. *Photochem. Photobiol. Sci., 1,* 741–748.

94. Mamada, A., Tanaka, T., Kungwatchakun, D., & Irie, M., (1990). Photoinduced phase transition of gels. *Macromolecules, 23,* 1517–1519.

95. Jiang, J. Q., Tong, X., & Zhao, Y., (2005). A new design for light-breakable polymer micelles. *J. Am. Chem. Soc., 127,* 8290–8291.

96. Rijcken, C. J., Soga, O., Hennink, W. E., & Van Nostrum, C. F., (2007). Triggered destabilization of polymeric micelles and vesicles by changing polymers polarity: An attractive tool for drug delivery. *J. Control. Release, 120,* 131–148.

97. Silva, J. N., Filipe, P., Morlière, P., Mazière, J. C., Freitas, J. P., Gomes, M. M., & Santus, R., (2008). Photodynamic therapy: Dermatology and ophthalmology as main fields of current applications in clinic. *Biomed. Mater. Eng., 18*(1–5), 311–327.

98. Verestiuc, L., Nastasescu, O., Barbu, E., Sarvaiya, I., Green, K. L., & Tsibouklis, J., (2006). Functionalized chitosan/NIPAM (HEMA) hybrid polymer networks as inserts for ocular drug delivery: Synthesis, *in vitro* assessment, and *in vivo* evaluation. *J. Biomed. Mater. Res. A., 77*(4), 721–735.

99. Recum, H. V., Kikuchi, A., Yamato, M., Sakurai, Y., Okano, T., & Kim, S. W., (1999). Growth factor and matrix molecules preserve cell function on thermally responsive culture surfaces. *Tissue Eng., 5,* 251–265.

100. Galperin, A., Long, T. J., & Ratner, B. D., (2010). Degradable, thermo-sensitive poly(N-isopropyl acrylamide)-based scaffolds with controlled porosity for tissue engineering applications. *Biomacromolecules, 11,* 2581–2592.

101. Kang Derwent, J. J., & Mieler, W. F., (2008). Thermoresponsive hydrogels as a new ocular drug delivery platform to the posterior segment of the eye. *Trans. Am. Ophthalmol. Soc., 106,* 201–213.

102. Kumar, S., Haglund, B. O., & Himmelstein, K. J., (1994). *In situ*-forming gels for ophthalmic drug delivery. *J. Ocul. Pharmacol., 10,* 41–56.

103. Xie, B., Jin, L., Luo, Z., Yu, J., Shi, S., Zhang, Z., Shen, M., Chen, H., Li, X., & Song, Z., (2015). An injectable thermosensitive polymeric hydrogel for sustained release of Avastin® to treat posterior segment disease. *Int. J. Pharm., 490,* 375–383.

104. Ninawe, P. R., Hatziavramidis, D., & Parulekar, S. J., (2010). Delivery of drug macromolecules from thermally responsive gel implants to the posterior eye. *Chem. Eng. Sci., 65,* 5171–5177.

105. Tsai, T. S., Pillay, V., Choonara, Y. E., Toit, L. C., Modi, G., Naidoo, D., & Kumar, P., (2011). A polyvinyl alcohol-polyaniline based electro-conductive hydrogel for controlled stimuli-actuable release of indomethacin. *Polymers, 3,* 151–172.

106. Sultana, Y., Aqil, M., & Ali, A., (2006). Ion-activated, Gelrite®-based *in situ* ophthalmic gels of pefloxacin mesylate: Comparison with conventional eye drops. *Drug Delivery, 13,* 211–219.

107. Vodithala, S., Khatry, S., Shastri, N., & Sadanandam, M., (2010). Formulation and evaluation of ion activated ocular gels of ketorolac tromethamine. *Inter. Jour. of Curr. Pharm. Research, 2,* 31–38.

108. Hui, H. W., & Robinson, J. R., (1985). Ocular delivery of progesterone using a bioadhesive polymer. *Int. J. Pharm., 26,* 201–213.

109. Rozier, A., Mazuel, C., Grove, J., & Plazonnet, B., (1989). Gelrite: A novel ion activated *in situ* gelling effect bioavailability of timolol. *Int. J. Pharm., 57,* 161–168.

110. Sanzgiri, Y. D., Maschi, S., Crescenzi, V., Callengaro, L., Topp, E. M., & Stella, V. J., (1993). Gellan based system for ophthalmic sustained delivery of methyl prednisolone. *J. Control. Release, 26,* 195–201.

111. Maurice, D. M., & Srinivas, S. P., (1992). Use of flurometry in assessing the efficacy of a cation-sensitive gel as an ophthalmic vehicle: Comparison with scintigraphy. *J. Pharm. Sci., 81,* 615–619.

112. Cohen, S., & Lobel, E., (1997). A novel *in situ* forming ophthalmic drug delivery system from alginates undergoing gelation in the eye. *J. Control. Release, 44,* 201–208.

113. Lokhande, U. R., Gorde, V. D., Gadhave, M. V., Jadhav, S. L., & Gaikwad, D. D., (2012). Design and development of pH-triggered *in situ* gelling system of ciprofloxacin. *Int. Res. J. Pharm., 3*(5), 411–422.

114. Singh, V., Bushetti, S. S., Appala, R., Shareef, A., Imam, S., & Singh, M., (2010). Stimuli-sensitive hydrogels: A novel ophthalmic drug delivery system. *Indian J. Ophthalmol., 58*(6), 471–481.

115. Rathod, K. B., & Patel, M. B., (2014). Controlled release *in situ* gel of norfloxacin for ocular drug delivery. *Int. J. Pharm. Sci. Res., 5*(6), 2331–2336.

116. Shetty, N., & Charyulu, R. N., (2013). A study on stability and *in vivo* drug release of naphazoline and antazoline *in situ* gelling systems for ocular delivery. *Int. J. Pharm. Bio. Sci., 4*(1), 161–171.

117. Saxena, P., & Kushwaha, S. K., (2014). pH sensitive hydrogels of levofloxacin hemihydrates for ophthalmic drug delivery. *WJPR, 3*(2), 4001–4022.

118. Nayak, S., Sogali, B. S., & Thakur, R. S., (2012). Formulation and evaluation of pH triggered *in situ* ophthalmic gel of moxifloxacin hydrochloride. *Int. J. Pharm. Pharm. Sci., 4,* 451–459.

119. Preetha, J. P., Karthika, K., Rekha, N. R., & Elshafie, K., (2010). Formulation and evaluation of *in situ* ophthalmic gels of diclofenac sodium. *J. Chem. Pharm. Res., 2*(3), 521–535.

120. Patel, P. B., Shastri, D. H., Shelat, P. K., & Shukla, A. K., (2011). Development and evaluation of pH triggered *in-situ* ophthalmic gel formulation of ofloxacin. *Am. J. Pharm. Tech. Res., 1*(4), 431–445.

121. Nagalakshmi, S., Seshank, R. R., & Shanmuganathan, S., (2014). Formulation and evaluation of stimuli sensitive ph triggered *in-situ* gelling system of fluconazole in ocular drug delivery. *Int. J. Pharm. Sci. Res., 5,* 1331–1344.

122. Kanoujia, J., Sonker, K., Pandey, M., Kymonil, K. M., & Saraf, S. A., (2012). Formulation and characterization of a novel pH-triggered *in-situ* gelling ocular system containing gatifloxacin. *Int. Curr. Pharm. J., 1*(3), 41–49.

123. Patel, R., & Patel, K. P., (2010). Advances in novel parentral drug delivery systems. *Asian J. Pharm., 4,* 191–199.

124. Packhaeuser, C. B., Schnieders, J., Oster, C. G., & Kissel, T., (2004). *In situ* forming parenteral drug delivery systems: An overview. *Eur. J. Pharm. Biopharm., 58,* 441–455.

125. Sleem, M., Alam, A., Ahmed, S., Iqbal, M., & Sultana, S., (1999). Tephrosia purpurea ameliorates benzoyl peroxide induced cutaneous toxicity in mice: Diminution of oxidative stress. *Pharmacy and Pharmacological Communication, 5,* 455–461.

126. Dunn, R. L., (2003). The atrigel drug delivery system. In: Rathbone, M. J., Hadgraft, J., & Roberts, M., (eds.), *Modified-Release Drug Delivery Technology* (1st edn., p. 647). Marcel Dekker: New York.

127. Ravivarapu, H. B., Moyer, K. L., & Dunn, R. L., (2000). Sustained activity and release of leuprolide acetate from an *in situ* forming polymeric implant. *AAPS Pharm. Sci. Tech., 1,* 1–8.

128. Radomsky, M. L., Brouwer, G., Floy, B. J., Loury, D. J., Chu, F., Tipton, A. J., & Sanders, L. M., (1993). The controlled release of Ganirelix from the Atrigel™ injectable implant system, *Proc. Int. Symp. Control. Rel. Bioact. Mater, 20,* 458–459.

129. Lambert, W. J., & Peck, K. D., (1995). Development of an *in situ* forming biodegradable poly-lactide-coglycolide system for the controlled release of proteins. *J. Control. Release, 33,* 181–195.

130. Shively, M. L., Bennett, A. T., Coonts, B. A., Renner, W. D., & Southard, J. L., (1995). Physico-chemical characterization of a polymeric injectable implant delivery system. *J. Control. Release, 33,* 231–243.

131. Chandrashekar, B. L., Zhou, M., Jarr, E. M., & Dunn, R. L., (2000). *Inventors*. Atrix Laboratories Inc., assignee, US6143314.

132. Brodbeck, K. J., Desnoyer, J. R., & Mchugh, A. J., (1999). Phase inversion dynamics of PLGA solutions related to drug delivery: Part II. The role of solution thermodynamics and bath-side mass transfer. *J. Control. Release, 62,* 331–344.

133. Graham, P. D., Brodbeck, K. J., & Mchugh, A. J., (1999). Phase inversion dynamics of PLGA solutions related to drug delivery. *J. Control. Release, 58,* 231–245.

134. Gentner, G., (2001). Biodegradable block copolymers for delivery of proteins and water-insoluble drugs. *J. Control. Release, 72,* 201–215.
135. Bezwada, R. S., (1995). *Liquid Copolymers of Epsilon-Caprolactone and Lactide.* U.S. Patent 5,442,033.
136. Shinoda, H., Ajioka, M., & Chida, K., (1998). *Bioabsorbable Polymer and Process for Preparing the Same.* U.S. Patent 5,747,637.
137. Moreau, M., Schneider, M., Boisramc, B., & Gurny, R., (2002). Controlled delivery of metoclopramide using on injectable semisolid poly(orthoester) for veterinary application. *Int. J. Pharm., 248,* 31–37.
138. Einmahl, S., Behar-Cohen, F., Tabatabay, C., Hermies, F. D., Chauvaud, D., Heller, J., & Gurny, R. A., (2000). Viscous bioerodible poly(orthoester) as a new biomaterial for intraocular application. *Journal Biomed Mater Research, 50,* 561–577.
139. Holland, S. J., Tighe, B. J., & Gould, P. L., (1986). Polymers for biodegradable medical devices. 1. The potential of polyesters as controlled macromolecular release systems. *J. Control. Release, 4,* 151–180.
140. Deshpande, A. A., Heller, J., & Gurny, R., (1998). Bioerodible polymers for ocular drug delivery. *Crit. Rev. Ther. Drug Carrier Syst., 15,* 381–420.
141. Mali, M., & Hajare, A., (2009). Systems for sustained ocular drug delivery: A review of stimuli-sensitive *in situ* gel-forming systems. *Euro J. of Parent and Pharm. Sci., 14,* 71–83.
142. Zhang, X., Jackson, J. K., Wang, W., Min, W., Cruz, T., Hunter, W. L., & Burt, H. M., (1996). Development of biodegradable polymeric paste formulation for taxol: An *in-vivo* and *in-vitro* study. *Int. J. Pharm., 137,* 191–208.
143. Karewicz, A., Zasada, K., Szczubiałka, K., Zapotoczny, S., Lach, R., & Nowakowska, M., (2010). "Smart" alginate-hydroxypropylcellulose microbeads for controlled release of heparin. *Int. J. Pharm., 385,* 161–169.
144. Kono, K., Ozawa, T., Yoshida, T., Ozaki, F., Ishizaka, Y., Maruyama, K., Kojima, C., Harada, A., & Aoshima, S., (2010). Highly temperature-sensitive liposomes based on a thermosensitive block copolymer for tumor-specific chemotherapy. *Biomaterials, 31,* 7091–7105.
145. Suisha, F., Kawasaki, N., Miyazaki, S., Shirakawa, M., Yamatoya, K., Sasaki, M., & Attwood, D., (1998). Xyloglucan gels as sustained release vehicles for the intraperitoneal administration of mitomycin C. *Int. J. Pharm., 172,* 21–32.
146. Ding, C., Zhao, L., Liu, F., Cheng, J., Gu, J., Dan, S., Liu, C., Qu, X., & Yang, Z., (2010). Dually responsive injectable hydrogel prepared by *in situ* cross-linking of glycol chitosan and benzaldehyde-capped PEO-PPO-PEO. *Biomacromolecules, 11,* 1043–1051.
147. Salehi, R., Arsalani, N., Davaran, S., & Entezami, A. A., (2009). Synthesis and characterization of thermosensitive and pH-sensitive poly(N-isopropylacrylamide-acrylamide-vinylpyrrolidone) for use in controlled release of naltrexone. *J. Biomed. Mater. Res., 89A,* 911–928.
148. Zhou, L., Cheng, R., Tao, H., Ma, S., Guo, W., Meng, F., Liu, H., Liu, Z., & Zhong, Z., (2011). Endosomal pH-Activatable poly(ethylene oxide)-graft-doxorubicin prodrugs: Synthesis, drug release, and biodistribution in tumor-bearing mice. *Biomacromolecules, 12,* 1461–1467.

149. Ghosn, B., Singh, A., Li, M., Vlassov, A. V., Burnett, C., Puri, N., & Roy, K., (2010). Efficient gene silencing in lungs and liver using imidazole-modified chitosan as a nanocarrier for small interfering RNA. *Oligonucleotides, 20,* 161–172.

150. Lai, T. C., Bae, Y., Yoshida, T., Kataoka, K., & Kwon, G. S., (2010). pH-sensitive multi-PEGylated block copolymer as a bioresponsive pDNA delivery vector. *Pharm. Res., 27,* 2260–2273.

151. Yu, C., Gao, C., Lü, S., Chen, C., Yang, J., Di, X., & Liu, M., (2014). Facile preparation of pH-sensitive micelles self-assembled from amphiphilic chondroitin sulfate-histamine conjugate for triggered intracellular drug release. *Colloids Surf. B Biointerfaces, 115,* 331–339.

152. Lee, E. S., Gao, Z., Kim, D., Park, K., Kwon, I. C., & Bae, Y. H., (2008). Super pH-sensitive multifunctional polymeric micelle for tumor pHe specific TAT exposure and multidrug resistance. *J. Control. Release, 129,* 221–236.

153. Wu, X. L., Kim, J. H., Koo, H., Bae, S. M., Shin, H., Kim, M. S., et al., (2010). Tumor-targeting peptide conjugated pH-responsive micelles as a potential drug carrier for cancer therapy. *Bioconjugate Chem., 21*(2), 201–213.

154. Etrych, T., Sírová, M., Starovoytova, L., Ríhová, B., Ulbrich, K., Etrych, T., & Sirova, M., (2010). HPMA copolymer conjugates of paclitaxel and docetaxel with pH-controlled drug release. *Mol. Pharmaceutics, 7,* 1015–1026.

155. Bae, Y., Nishiyama, N., & Kataoka, K., (2007). *In vivo* antitumor activity of the folate-conjugated pH-sensitive polymeric micelle selectively releasing adriamycin in the intracellular acidic compartments. *Bioconjugate Chem., 18,* 1131–1139.

156. Cohen, J. A., Beaudette, T. T., Tseng, W. W., Bachelder, E. M., Mende, I., Engleman, E. G., & Fréchet, J. M., (2009). T-cell activation by antigen-loaded pH-sensitive hydrogel particles *in vivo*: The effect of particle size. *Bioconjugate Chem., 20,* 111–119.

157. Shin, J., Shum, P., & Thompson, D. H., (2003). Acid-triggered release via dePE-Gylation of DOPE liposomes containing acid-labile vinyl ether PEG-lipids. *J. Control. Release, 91,* 181–200.

158. Lin, C., Zhong, Z., Lok, M. C., Jiang, X., Hennink, W. E., Feijen, J., & Engbersen, J. F., (2007). Novel bioreducible poly(amido amine)s for highly efficient gene delivery. *Bioconjugate Chem., 18,* 131–145.

159. Xu, J., Gao, F., Li, L., Ma, H. L., Fan, Y., Liu, W., Guo, S., Zhao, X., & Wang, H., (2013). Gelatin-mesoporous silica nanoparticles as matrix metalloproteinases-degradable drug delivery systems *in vivo*. *Microporous Mesoporous Mater, 182,* 161–172.

160. Hua, M. Y., Liu, H. L., Yang, H. W., Chen, P. Y., Tsai, R. Y., Huang, C. Y., et al., (2011). The effectiveness of a magnetic nanoparticle-based delivery system for BCNU in the treatment of gliomas. *Biomaterials, 32,* 511–527.

161. Tung, W. L., Hu, S. H., & Liu, D. M., (2011). Synthesis of nanocarriers with remote magnetic drug release control and enhanced drug delivery for intracellular targeting of cancer cells. *Acta Biomater, 7,* 2873–2882.

162. Schroeder, A., Honen, R., Turjeman, K., Gabizon, A., Kost, J., & Barenholz, Y., (2009). Ultrasound triggered release of cisplatin from liposomes in murine tumors. *J. Control. Release, 137,* 61–68.

163. Nombona, N., Maduray, K., Antunes, E., Karsten, A., & Nyokong, T., (2012). Synthesis of phthalocyanine conjugates with gold nanoparticles and liposomes for photodynamic therapy. *J. Photochem. Photobiol. B., 107,* 31–44.

164. Jeong, S. Y., Yi, S. L., Lim, S. K., Park, S. J., Jung, J., Woo, H. N., et al., (2009). Enhancement of radiotherapeutic effectiveness by temperature-sensitive liposomal 1-methylxanthine. *Int. J. Pharm., 372*, 131–139.

165. Zhang, T., Sturgis, T. F., & Youan, B. B. C., (2011). pH-responsive nanoparticles releasing tenofovir intended for the prevention of HIV transmission. *Eur. J. Pharm. Biopharm., 79,* 521–536.

166. Huynh, D. P., Im, G. J., Chae, S. Y., Lee, K. C., & Lee, D. S., (2009). Controlled release of insulin from pH/temperature-sensitive injectable pentablock copolymer hydrogel. *J. Control. Release, 137*, 21–24.

167. Oberoi, H. S., Laquer, F. C., Marky, L. A., Kabanov, A. V., & Bronich, T. K., (2011). Core cross-linked block ionomer micelles as pH-responsive carriers for cis-diamminedichloroplatinum(II). *J. Control. Release, 153*, 61–72.

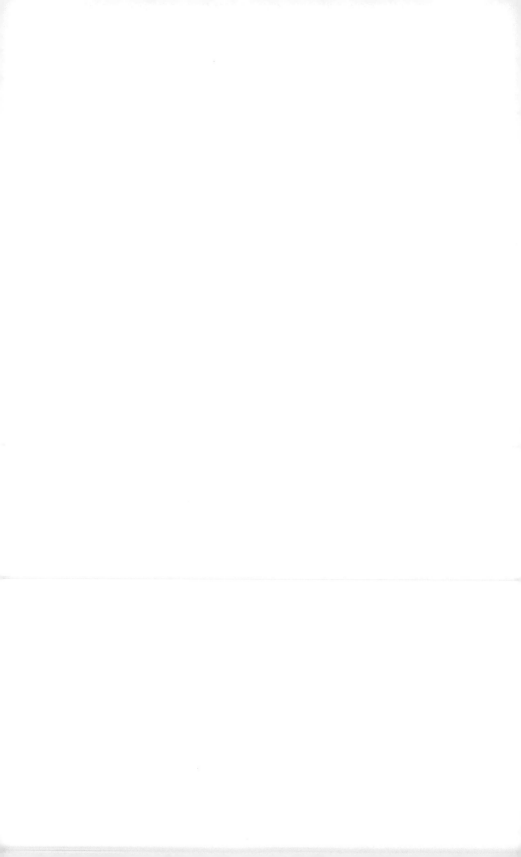

# CHAPTER 4

# Application of Bioresponsive Polymers in Gene Delivery

TAMGUE SERGES WILLIAM, DRASHTI PATHAK, and DEEPA H. PATEL

*Department of Pharmaceutics, Parul Institute of Pharmacy and Research, Faculty of Pharmacy, Parul University, P.O. Limda, Ta. Waghodia, Vadodara–391760, Gujarat, India, Phone: 02668-260287, Fax: 02668-260201, E-mails: deepaben.patel@paruluniversity.ac.in, Pateldeepa18@yahoo.com (D. H. Patel)*

## 4.1 INTRODUCTION

Some diseases such as cancers or genetic diseases are caused by the absence or mutation of a particular gene leading to the production of defective proteins which will entail to the pathological known state. Many approaches for the treatment of such diseases have been investigated and employed among which gene therapy is the most promising. Gene therapy is a medical strategy based on the delivery of genes as therapeutic agents for the correction of a mutation, prevention, or treatment of pathology. It involves the insertion of the therapeutic nucleic acid fragment into target cells to cause their expression in the desired proteins. By this approach replacement of damaged genes or inhibition of expression of undesired genes can be achieved.

The therapeutic nucleic acid administration can be achieved in two ways: *ex vivo* and *in vivo*. The *ex vivo* method is based on the cell transplantation technique and involves the harvesting of targeted cells from the patient, then the cultivation and transfer of gene *in vitro*. In the end, the cells thus transfected are reimplanted. Nevertheless, this approach is difficult to implement and is of the high cost.

The *in vivo* method involves the direct introduction of gene or vector containing gene into the target organ of the patient by diverse administration routes or methods (systemic or local). This later is generally limited by physiological barriers and required many injections.

A different system of delivery is available and involves the applications of non-viral plasmid (pDNA) expression vectors; small interfering RNA (siRNA); single-stranded antisense oligodeoxynucleotides (ODNs) or an immune system stimulating nucleic acid. An ideal gene therapy delivery system would be injectable, targetable to specific sites *in vivo*, regulatable, and able to maintain long-term gene expression, and be non-immunogenic. The difficulty in this area would be to design the perfect vector integrating all the parameters required for the efficient expression of the gene. The particles containing DNA must successively process in three major steps (1) to attach with cells, (2) enter the cytoplasm of the cell, by direct fusion with the plasma membrane, either after rupture of intracellular vesicles (endosomes, lysosomes), (3) allow the entry of DNA into cell nucleus [1].

To achieve it, two major systems have been used for a while; viral and non-viral mediated systems. Viral mediated system is based on the uses of various viruses as carriers to insert genetic sequence into the host cell. The viruses have the ability to cross cell barriers and insert their genetic material inside the host cell. Some examples of viruses used include retroviral vectors as human immunodeficiency virus (HIV), adeno-associated virus (AAV), herpes simplex virus (HSV), Epstein Barr virus (EBV), adenoviruses (AV), hepatitis B virus (HBV), Moloney murine leukemia virus (MoMLV). Creating a viral vector involves producing a recombinant virus lacking replication but maintaining its ability to infect cells [2]. These viral vectors are generally efficient tools of transfection but they possess some disadvantages like high production cost, difficulty in targeting to certain cells, size limitation of DNA constructs safety concerns, immunogenic reactions, toxic side effects, and possibility of triggering oncogenes, possibility to revert back or to retain an infectious form. The tentative to overcome such drawbacks has opened the gates to exploration of alternatives methods, which are non-viral mediated systems. In the recent past, non-viral mediated gene transfer has gained much more interest as potentially safe and effective methods to transfer genes in a wide range of genetic disease.

Non-viral mediated gene therapy includes delivery of naked DNA; encapsulated DNA; artificial chromosomes and self-assembled complexes. It is based on many different strategies to achieve safe and effective gene delivery. Carrier's systems used are diverse such as liposomes, cationic lipids, oligonucleotide carrier system, nanostructures, synthetic retransposon vectors, peptides, and polymers based gene delivery systems, electroporation, and miscellaneous approaches (engineered stem cells, transgenic approach, targeted modification of human genes).

Among these diverse techniques, polymers based approach represents one of the best ways to avoid the immunogenicity and expensive production costs encountered with viral vectors, because it allows a wide range of chemical modifications of the polymeric carrier. Furthermore, the fact that nucleic acid themselves are negatively charged polymers which can easily form complexes with polycations molecules (polyplexes) have made polymers a much more interesting and effective tools for gene transfer.

Numerous non-viral mediated gene delivery systems based cationic polymers have been developed to overcome safety issues related to viral vectors. Remarkable efforts have been made to optimize gene delivery polymers *in vitro*. *In vitro*, they have shown good transfection efficiency limited only by the presence of serum proteins which are the main obstacles. While, *in vivo* application may require different vector features because to achieve the goal of gene therapy, numerous hurdles that reduce their efficiency have to be surmounted by complexes, such as extracellular circulation, lack of tissue targeting, limited endocytic capacity of cells, lack of specificity between ligands and cellular receptors, toxicity, and immunogenicity of targeting moiety, enzymatic degradation, and destruction of endocyted complexes, insufficient release of the loaded content. These obstacles led to a difficult correlation between *in vitro* and *in vivo* results.

To overcome these problems and achieve a localized and controlled release of therapeutic nucleic acids at the targeted site, as well as to enhance its stability and to assure its efficient and safe delivery from the polymeric system, many modifications have been made over polymers. Numerous polymers design strategies have been reported, for the most focused on chemical modifications. The elaboration of smart polyplexes based on bioresponsive polymers, highly efficient and compatible showed a great promises and interest, in application for gene delivery.

## 4.2    BIORESPONSIVE POLYMERS FOR GENE DELIVERY

### 4.2.1    *BIORESPONSIVE POLYMERS IN NON-VIRAL DELIVERY SYSTEMS*

One of the major drawbacks of conventional lipoplexes and polyplexes is the insufficient stability of the complexes outside balanced by too high stability inside the cell, due to the strong condensation of nucleic acids with polymers. That high stability limits the release of the biotherapeutic agent at the target site. It has been thought that polymers could be designed to respond to microenvironmental parameters with changes in their physicochemical properties, enabling them to perform individual and dependent delivery tasks. Thus, bioresponsive polymers have been developed as a logical solution to this polyplexes dilemma. Bioresponsive polymers are defined as polymers that will undergo reversible, physical, or chemical changes in response to external modifications of the environmental conditions. These polymers will react to Subtle changes in the environment such as, unusual acidic conditions, temperature change in the target tissue, concentration modifications of some protein (glutathione (GSH) inside cells compare to plasma) of some biological molecules, light, and some others, which are considered as the trigger or biostimuli that induce the release of the therapeutic agent at the targeted site.

#### 4.2.1.1    *pH-RESPONSIVE POLYMERS FOR GENE DELIVERY*

A pH-responsive system exploits the pH variation in the body. There are distinct pH gradients in different cells, tissues, and even different cellular compartments, as well as under different physiological and pathological conditions. Genes delivery based pH-responsive polymers involves mainly two distinct mechanisms:

The first one which is a protonation/deprotonation event, lean on the variation of acidic conditions between the normal physiological environment pH 7.4 and those of endosome and lysosome which drop to 5.5 and 4.5, respectively [6]. The same as many microorganisms like viruses and bacteria which can naturally destabilize endosome through their pH-responsive surface proteins, pH-responsive polymers exploit this pH gradient between endosome and cytoplasm to protonate and then promote efficient transit in the cytosol and/or nuclei. According to the fact that polymers are mainly internalized through the endosome route, pH-responsive polymers are the most used synthetic vectors for the delivery of therapeutic nucleic acid.

The second mechanism is based on the destabilization or degradation of a pH-sensitive linker, which conjugates to the carrier and the cargo. The conjugated moieties can be acid labile-linkers such as acetal, ketal, hydrazone, vinyl ester, and orthoester groups which will affect the conformation or hydrophilicity of the polymer after a pH change.

### 4.2.1.1.1   pH-Responsive Protonatable Structures

They undergo the mechanism of protonation when they are present in acidic medium. These polymers can be from natural origin (albumin, gelatin, chitosan (CS)), or synthetics examples include poly(L-lysine) (PLL), polyethyleneimine (PEI), Polyaspartamide (PAsp) and poly(N,N-diakylaminoethyl methacrylates) (PDAAEMA).

### 4.2.1.1.2   Polyethylenimine (PEI)-Based Polyplexes

Polyethylenimines (PEIs) have become one of the most highly used nowadays due to its ability to effectively condense nucleic acids into nanoscale particles. It was first introduced by Behr et al. [7, 8]. Many types have been synthesized but the most important among them is linear (22 kDa) and marketed under the name ExGenTM500 and recently jetPEI™ [9].

They present very good characteristics:

- They are stable because of an efficient condensation of DNA;
- They have a small size (around 100 nm);
- They possess a high buffering capacity which allows them to cover a wide pH range, this capacity is called 'proton sponge effect' and it is due to important numbers of amino groups (primary, secondary, and tertiary) for branched polymers;
- They provide a good protection against nucleases degradation and a high efficient transfection capacity *in vitro* as well as *in vivo* possibly explained by the 'sponge effect' [7].

Endosomal escape is known as one of the most important steps which can lead in a limited and non-efficient transfection of polyplexes. PEI-based polyplexes easily promote endosomal release due to their well-known high buffering properties which enable endosomal pH-responsive protonation. The facilitated protonation of PEI involves an extensive influx of protons ions followed by chloride ions, which induce the increase of the osmotic pressure within the endosome, resulting to the endosomal disruption and the subsequent release of entrapped substances to the cytosol. That phenomenon is called "proton sponge" effect [8].

The amino groups of PEI can protonate at the endosomal pH 5.0 with a protonation degree of around 40% whilst it is only 20–25% at the physiological pH of 7.4 [10].

Compared with the control polyplex constituted of PLys with high pKa amines, PEI based Polyplexes are demonstrated to be of much higher efficiency in the gene expression of pDNA in cultured cells [8].

PEI because of its lack of biodegradability, can often induce cytotoxicity and immunogenicity, it is the reason why various modification strategies have been attempted on its structure like addition of bioresponsive linkers (disulfide bonds, acid-labile bonds), and shielding with PEG.

In 2016, Guan et al. [11] have developed an ultrasensitive pH triggered charge/size dual-rebound gene delivery system for efficient tumor treatment. The system is based on PEI and poly-l-glutamate (PLG) complexes with the negatively charged pDNA. The complex was later tightened with aldehyde modified PEG groups at both termini; and it showed stability at neutral pH but was cleavable in tumor extracellular pH.

Some advantages have been reported from this gene delivery system such as decrease of cytotoxicity, increase of stability, prolongation of circulation due to PEG shielding which could be rapidly removed due to acidic pH when reaching tumors [11] (Figure 4.1).

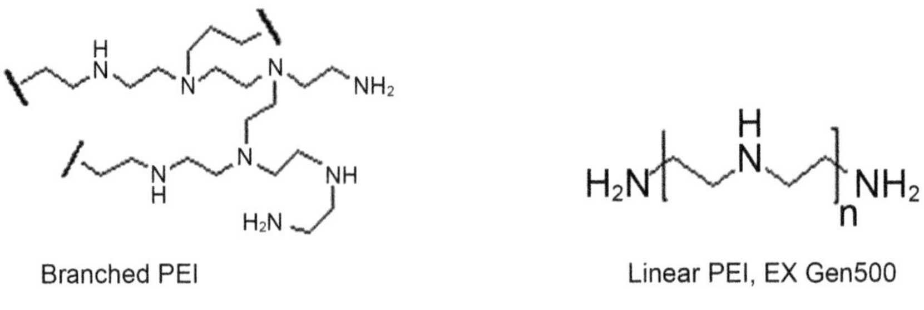

Branched PEI                                    Linear PEI, EX Gen500

**FIGURE 4.1**  Structures of polyethylenimines polymers.

### 4.2.1.1.3    *Polyaspartamide Based Polyplexes: PAsp(DET)*

It is a polyaspartamide derivative which consists of diaminomethane groups with two distinct PKa 6.2 and 8.9 [12]. As same as for PEI, PAsp (DET) undergoes a pH-responsive protonation in endosome. Its diaminomethane moiety in the side chain will change from a normal monoprotonated state at 7.4 pH to a diprotonated state at the endosomal acidic pH 5.0. Thus leading to a strong reaction with membrane and destabilization of this latter because of the high cationic charge density formed.

However, unlike PEI based polyplexes, polyaspartamide based polyplexes are biodegradable. Their main chain could degrade through self-catalytic reaction at neutral pH, at 37°C. By doing so, they provide good transfection efficiency without or with less sides effects. Thus, PAsp (DET) selectively destabilized the cellular membrane at the late endosomal pH for endosomal escape with minimal cytotoxicity.

Kataoka et al. (2008) have developed ternary polyplexes based on PAsp (DET) that express negative charge at extracellular pH, but turn positive to destabilize endosome at endosomal pH. The polyplexes showed good transfection efficiency and lower cytotoxicity against HUVEC cells [13].

PAsp (DET) has been successfully used to prepare polyplex micelles with pDNA and mRNA and its utility have been demonstrated for animal disease models. Thus Ge et al. (2014) have developed polyplexes micelles based polyethyleneglycol Polyaspartamide (PAsp) PEG-b-PAsp(DET) and pDNA encoding sFlt-1 that showed important inhibition of the growing of subcutaneous pancreatic tumor in mice [14, 15] (Figure 4.2).

### 4.2.1.1.4    *Poly(Dimethyl Amino Ethyl Methacrylate) (PDMAEMA)*

PDMAEMA is a proton sponge or membrane disruptive polymer like PEI; it also possesses the good buffering capacity and has been developed and shown for the first time by Hennink *et al.* (1996).

PDMEAMA is a polymer that can be easily obtained by radical polymerization of the corresponding vinyl monomer, or also by anionic and controlled radical polymerization techniques such as atom transfer radical polymerization (ATRP) or reversible addition-fragmentation transfer (RAFT). It contains only tertiary amines and can reach 90% of PEI (branched PEI 25kDa) transfection efficiency [16]. It shows ionization behavior at low pH and can be readily reversed by the pH change.

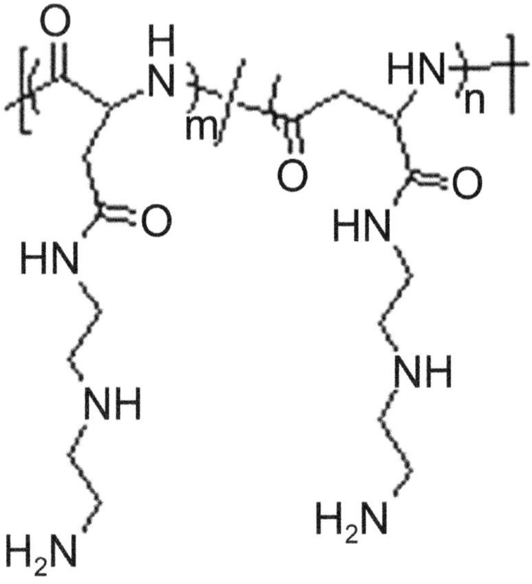

**FIGURE 4.2**    Schematic structure of PAsp (DET).

Samsonova and co-workers [17] have prepared different molecular weight PDMAEMAs and have reported that its transfection efficiency and toxicity are correlated to its molecular weight. Thus, PDMAEMA has been the subject of many studies, which attempted to prepare the ideal carrier showing acceptable efficiency and toxicity.

Bruns et al. [18] have developed micelles constituted of amphiphilic poly(dimethylsiloxane)-b-poly(2-(dimethylamino)ethyl methacrylate) (PDMS-b-PDMAEMA) block copolymers. The polymer self-assembled into micelles in aqueous solution and show the capacity to release its contents in response to a pH change from 7.4 to 5.5 micelle containing longer PDMAEMA block length showed increased toxicity (Figure 4.3).

### 4.2.1.1.5   pH-Responsive-Based Degradable Structure

**1. Polycations Conjugated With Acid-Labile Linkage:** The acid-labile linkage is a moiety added to the polymeric chain which will be hydrolyzed as soon as it is in the contact of acidic pH. Thus after the internalization of polyplexes, the acid-labile linkage could be hydrolyzed at the endosomal pH and then the inner positive charged polyplex will be unmasked or exposed, which will lead to membrane disruption by the same to endosomal escape [19].

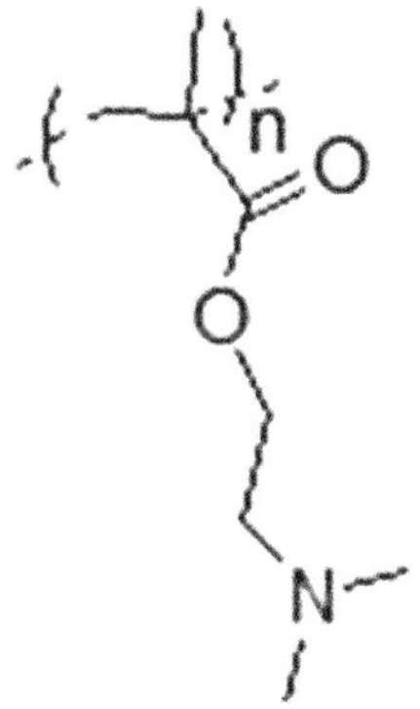

**FIGURE 4.3**    Structure of PDMAEMA polymer.

Acid labile-linkers such as acetal, ketal, hydrazone, imine, orthoester, and maleic acid are the most used to achieve pH-responsive degradability of the polyplexes. The linkers are fixed either into the main chain or the side chain of the polycation, enabling the site-specific release of the therapeutic nucleic acids with a low toxicity.

Li and co-workers have prepared polyplex nanoparticles based on plasmid DNA condensed with an acid-labile block copolymer consisting of poly(ethylene glycol) PEG and Poly(2-(dimethylamino)ethyl methacrylate)(PEG-a-PDMAEMA) segments linked together by a cyclic orthoester linkage (acid cleavable end group). The block copolymer was pre-synthesized by ARTF of DMAEMA using a PEG macroinitiator. The acid-labile polymer has shown a great increase in transfection efficiency after 6 hours at pH 5.0 compared to the stable PEG-b-PDMAEMA polymer [20].

### 4.2.1.1.6    *Maleic Acid Amide (MAA) Derivatives*

MAA based polycation is obtained by a reaction between primary amine and maleic anhydride which leads to the formation of a polyanion. This latter will show a relative stability at neutral pH 7.4 but will be hydrolyzed as soon as it will be in contact with acidic pH. This will lead to the degradation of MAA and exposure of primary amines, thereby regenerating the parent polycation. Some studies reported that, in general the hydrolysis kinetics of MAA derivatives is related to the nature of the substituted groups at the double bonds of MAA moiety [21, 22]. Rozema and collaborators [22] have reported a faster hydrolysis from a MAA from dimethyl maleic acid derivatives that showed a degradation half-life of 1.5 min, compared to a MAA from monomethyl maleic acid derivatives (citraconic acid amide) which showed degradation half time of 300 min.

Many pH-responsive polymers using MAA derivatives as cellular internalization and endosomal escape enhancer have been developed, relying on the unique capacity of MAA and its derivatives to provide a negatively charged complex from a polycation, it is called charge conversion property (CCP) which afford interesting polymers designs.

pDNA/pLys polyplex, have been decorated at its surface by MAA-modified Pasp(DET) moiety, the decorated polyplex showed an enhanced endosomal escape, which enables it to exhibit 10-fold higher gene expression efficiency and lower cytotoxicity against human umbilical vein endothelial cells (HUVECs) compared with normal control polyplex [23]. For an efficient siRNA delivery into the cytosol of the target cell and a lower immunogenicity, Takemoto et al. have developed an acidic pH-responsive siRNA conjugate with multifunctionalities such as reversible carrier stability, endosomal escapability, and mono-siRNA release ability. The siRNA conjugate was constituted of a maleic acid amide (MAA) as acid-labile anionic moiety that links siRNA to an endosomal disrupting polycation (Pasp(DET) [24].

### 4.2.1.1.7  Poly(β-Aminoesters) (PAE)

pH-sensitive poly(β-amino ester) (PAE) was first reported by Langer et al. as a gene carrier, they are biodegradable polycations obtained by the Michael addition of either primary or bis-secondary amines monomers to diacrylate esters [25, 26]. Such biodegradable polycations degrades into nontoxic metabolites, it include poly-(l-lactide-co-l-lysine), poly(serine ester), poly(4-hydroxyl-proline ester), poly[α-(4-aminobutyl)-l-glycolic acid], hyperbranched poly(amino ester), which has showed good capacities to condense plasmid DNA and mediate gene transfer [25].

Amine-containing hydrolyzable polyesters possess the advantages of ease of synthesis by simple and inexpensive polymerization and have a lower cytotoxicity than poly-l-lysine (PLL) and Polyethylenimine (PEI) due to their biodegradability.

### 4.2.1.1.8    PAMAM Esters

Nam et al. (2008) have synthesized a biodegradable polycationic polyamidoamine (PAMAM) ester that containing arginine (e-PAM-R) or lysine (e-PAM-K) at the peripheral end of PAMAM-OH dendrimers through ester bonds linkages. They substituted the amide bonds of an arginine modified PAMAM dendrimers (PAMAM-R) by esters bonds for enhancement of degradation speed while conserving the remarkable transfection efficiency and the low toxicity.

More than 50% of grafted amino acids were degraded at physiological conditions (pH 7.4, 37°C) within 5h, with a relative stability of polyplexes at endosomal pH. The PAMAM esters derivatives have reported excellent buffering capacity and the arginine conjugated showed a better transfection efficiency than the branched PEI (25 kDa) [27] (Table 4.1).

### 4.2.1.2    THERMORESPONSIVE POLYMERS FOR GENE DELIVERY

Body temperature variation in some pathological conditions or diseases, especially in the tumor microenvironment (TME) (increase by 1–2°C), has been widely exploited as physical stimuli to serve as a trigger for the delivery of therapeutic agents in bioresponsive drug delivery system. Temperature-sensitive polymers have a specific critical solution temperature which can be set by upper critical solution temperature (UCST) that becomes soluble after heating or lower critical solution temperature (LCST) that becomes insoluble after heating. With small temperature changes around this temperature, there is a change in the hydrophilic and hydrophobic interactions between the polymer and the aqueous medium. That interaction changing leads to an architectural or structural modification of the polymer, which can be manifested by an expansion (swollen networks), or a collapse (collapsed network) [33].

Most of the thermosensitive polymers have a LCST and consist of a cationic segment, such as polyarginine, PEI, PLL, CS, and a thermosensitive polymer, such as PNIPAAM or its derivatives [34]. Poly(N-isopropyl acrylamide) (poly(NIPAAM)) is one of the most studied and

TABLE 4.1 Examples of pH-Sensitive Delivery Systems for Therapeutic Gene Delivery

| Polymer Used | pH-Responsive Moiety/Mechanism | Therapeutic Agent Used | References |
|---|---|---|---|
| Poly(ethylene imine) PEI | Proton sponge effect | pDNA siRNA | [59, 61] |
| Polyethylene glycol-b-polyaspartamide diamioethane (PEG-b-PAsp(DET)) | PAsp(DET) protonation | pDNA | [14, 15] |
| poly(dimethysiloxane)-b-poly(2-(dimethylamino)ethyl methacrylate) (PDMS-b-PDMAEMA) | PDMEAMA protonation | DOX | [18] |
| EGF-PEG (epidermal growth factor-poly ethylene glycol) conjugated PEI or PLL | Acidic environment labile; cell-targeting delivery | pDNA | [60] |
| Lipid-coated PBAE nanoparticles | Poly β amino ester (PBAE) | mRNA | [29] |
| Poly(2(dimethylamino)ethyl methacrylate)-blockpoly(2(diisopropylamino)ethyl methacrylate) (PDMA-b-PDPA) micelles | PDPA protonation | SiRNA | [30] |
| Poly(dimethylamino-ethyl methacrylate-block-butyl methacrylate) copolymers (pDbB) (micelle) | Poly(styrene-alt-maleic anhydride) (pSMA) | siRNA | [31] |
| PEI and hexanoate-PEI | Arginine rich peptides | pDNA, siRNA | [32] |
| PLL complexes with Lac-PEG-siRNA conjugate | Acid labile; cell targeting | siRNA | [69, 70] |
| dimethyl maleic anhydride-modified melittin (DMMAn-melconjugated PEG-PLL) | Acid sensitive | pDNA siRNA | [59, 71] |

used among thermoresponsive polymers. Its critical temperature close to body temperature, 32∘C (LCST), allows it to be highly soluble below this temperature and to change to an opaque gel form above this temperature [34]. Thus it has been intensively exploited and have undergone many different types of modifications (crosslinking, grafting of hydrophobic or hydrophilic moieties at its chain, copolymerization, etc.) to produce good materials with applications in different fields such as gene delivery, where the objectives are to obtain ideal thermosensitive carrier with best transfection abilities.

Based on these PNIPAAm modifications, various structures could have been realized such as hydrogels, micelles, polymersomes, nanoparticles, and others. Akimoto et al. [35] have prepared thermoresponsive micelles from pre-synthesized diblock copolymers comprising thermoresponsive segments of poly(N-isopropyl acrylamide-co-N,N-dimethylacrylamide) (P(IPAAm-co-DMAAm)) and hydrophobic segments of poly(D,L-lactide). The polymer showed a LCST of 40°C, below which the micelles were 25 nm size, and 600 nm when above the LCST. The thermoresponsive micelles showed time-dependent cellular uptake and lower toxicity above micellar LCST [35]. Lavigne et al. have developed a thermoresponsive polymer with LCST based polyethyleneimine (PEI) grafted with poly(N-isopropyl acrylamide) chains. This polymer displayed a coil globule transition when complexed to DNA. They showed that the applied changes in the vector configuration have enhanced transgene expression [36].

Mao et al. (2007) have successfully synthesized a thermoresponsive copolymer made up of trimethyl CS-grafted by poly(N-isopropyl acrylamide) (TMC-g-PNIPAAm) with an LCST (32°C). They showed a strong ability to combine with DNA at 40°C and optimized gene transfection efficiency comparable to Lipofectamine 2000, while no obvious cytotoxicity has been observed. Some other example of thermosensitive polymers which can be exploited as carriers for therapeutics nucleic acid delivery includes:

Copolymers of 2-(dimethylamino)ethyl methacrylate (DMAEMA) and N-isopropylacryl amide have been reported by Hinrichs et al. as an interesting thermosensitive carriers for DNA delivery. They have shown that the formation of stable (co)polymer/ plasmid complexes with a size of around 200 nm is a prerequisite for efficient transfection. Furthermore, the transfection efficiency and cytotoxicity strongly decreased with decreasing zeta potential [37].

Kurisawa et al. have synthesized a thermoresponsive copolymer (P (IP-8DA-11BM)), which is a poly(N-isopropylacrylamide (NIPAAM)-co-2-(dimethylamino) ethyl methacrylate (DMAEMA)-co-butylmethacrylate (BMA)) containing 8 mol% DMAEMA and 11 mol% BMA. The polymer has a low LCST of 21°C, form a stable complex with pDNA at 37°C and partially detached from the DNA at 20°C. It presents increased *in vitro* gene transfection efficiency at lower compare to other PDMAEMA homopolymers [38].

Bouchemal et al. have prepared a thermosensitive hydrogel based on amphiphilic triblock copolymers EOn-POm-EOn (assembled units of ethylene oxide and propylene oxide) pluronics® containing mucoadhesive hydroxyl propyl methylcellulose (HPMC). The hydrogel was incorporated with the peptide mini CD4 M48U1 used as an anti-HIV-1 molecule and has been shown as a promising tool which could be used as barrier against HIV infection, by preventing spreading of the virus [39] (Figure 4.4).

**FIGURE 4.4**    Schematic structures of thermosensitive polymers (a) PNIPAAM and (b) PDMAEMA [40].

### 4.2.1.3   REDUCTIVE ENVIRONMENT RESPONSIVE POLYMERS FOR GENE DELIVERY

This strategy exploits the reaction between the polymers and redox potential regulating agents such as GSH, thioredotoxin (Trx), cysteine (Cys) according to the site and the conditions.

It involves bioreducible cationic polymers or any other polymer that contains disulfide bond (-SS-) as bioreducible moiety in polymeric main chain or side chain. A Bioreducible polymer reacts to redox potential of the environment to undergo change in their physicochemical properties or conformation to release therapeutic nucleic acids.

Disulfide bond is relatively stable under the conditions of extracellular environment, but can be rapidly degraded inside the cells due to the presence of high amount of reducing agent such as GSH reductase and sulfhydryl component as GSH. Notably, disulfide linkages containing polyplexes are stable in the circulation but are dissociated in reducing endosomal environment. There is a huge difference in the GSH concentration between the intracellular medium and the bloodstream. Inside the cell, the GSH concentration is drastically elevated up to 1–10 mM, which is 50–1000 times higher than that in the bloodstream, generating highly reductive conditions [41, 42].

There are two main routes to generate bioreducible cationic polymers, which involve a reaction between two thiol groups for disulfide bond formation: the first one is the preparation of cationic polymer with pyridyldithio residue and then modification with a suitable thiol compound via an exchange reaction. The second one is the polyoxidation of dithiol-based monomers having amino groups, it is time consuming, and the compound undergoes oxidation of thiol groups by air.

### 4.2.1.3.1 Some Bioreducible Polymers as Gene Delivery Vectors

1. **Disulfide Bond Linked Synthetic Cationic Polymers (SS-PEI, SS-PLL, SS-PDMEAMA, SS-PAA):** Lee et al. have prepared a disulfide bond linked linear PEI(L-PEIS) which showed a good degradation capacity in cytosol of HeLa cells and an improvement of gene transfer. Such degradation could be accelerated in tumor cells due to the high concentration of GSH.

   Gaspar et al. [43] have synthesized a bioreducible PLA-PEI-SS triblock copolymer micelles for minicircle DNA and doxorubicin (DOX) co-delivery which is based on a poly(2-ethyl-2-oxazoline)-poly(L-lactide) grafted with bioreducible polyethyleneimine (PEOz-PLA-g-PEI-SS). The micelles have shown high gene expression compared with non-bioreducible carriers *in vitro* into tumor spheroids models. The co-delivery of mcDNA-Dox to B16F10-Luciferase tumor-bearing mice resulted in a reduction in tumor volume and cancer cell viability. That suggests the triblock copolymer as a promising tool for an efficient protection and co-delivery of DNA minicircles (mcDNA) and DOX, even if there is some cytotoxicity problem remaining [43].

Oupicky et al. [44] have developed a polymer-coated vector which manifests reversible lateral stabilization, thus promote extended circulation in the bloodstream and allow intracellular triggering of DNA release. The vector is based on a reductively degradable polycation RPC, coated, and surface cross-linked with copolymers of N-(2-hydroxypropyl) methacrylamide (PHPMA) which is responsible for the lateral stabilization. The RPC-based targeted vectors showed efficient intracellular activation, authenticating the concept that lateral stabilization introduced by surface coating with PHPMA can be reversed by intracellular reduction [44].

Another approach to obtain disulfide-based cationic polymers is to make an amine compound to react with a disulfide containing reagent such as cystamine bisacrylamide (CBA) through a Michael addition reaction. Via that method, Lin et al. (2007) have synthesized and evaluated a series of poly(amidoamine)s containing disulfide bonds (SS-PAAs) for gene delivery. The SS-PAA polymers showed good stability *in vitro* in a medium mimicking physiological conditions (pH 7.4, 150 mM PBS, 37°C) and was rapidly degraded in reductive environment such as the presence of 2.5 mM dithiothreitol (DTT) [45].

Kim et al. [46] have developed a compact nanosized polyplexes based on arginine-grafted poly(cystaminebisacrylamide diaminohexane) (ABP)-conjugated poly(amidoamine) (PAMAM) dendrimers (PAM-ABP) which showed a good transfection efficiency and a low cytotoxicity compared to bPEI 25kDa and Lipofectamine® 2000 [46].

2.  **Thiolation of Natural Polymers:** Natural polymers such as CS, gelatin, have also shown some interest because of their advantages such as biocompatibility, biodegradability, non-toxicity, and non-immunogenicity.

Stability for systemic gene delivery have been recently improved by using thiolated gelatin/DNA nanoparticles [47]. Thiolated gelatin nanoparticles has been reported to be a successful nanocarrier that respond to high concentration of GSH.

Kommareddy et al. have later developed a PEG modified thiolated gelatin (PEG-SHGel) nanoparticles encapsulated with a Reporter plasmid expressing enhanced green fluorescent protein (EGFP-N1). It showed good stability, the good release profile of encapsulated plasmid at various GSH concentrations (up to 5.0 mM GSH in PBS) and also the highest transfection efficiency of the reporter plasmid [47, 48].

The CS can be also a good alternative but present the disadvantage of a lower transfection efficiency compared with its synthetic polymer competitors, such as linear or branched PEI. Yu et al. have prepared complexes based Polypeptide/polysaccharide graft copolymers poly(L-lysine)-graft-chitosan (PLL-g-Chi). The copolymer showed gene transfection ability in HEK 293T cells superior to that of CS and dependent on the PLL grafting ratio [49] (Table 4.2).

### 4.2.1.4 ROS RESPONSIVE POLYMERS FOR NUCLEIC ACID DELIVERY

ROS are oxidants molecules derived from oxygen ($O_2$) which are either free radicals or molecular species capable of generating free radicals. They mainly comprise superoxide radicals, nitric oxide radicals, hydroxyl radical (OH), alkoxy radicals (RO), peroxyl radicals, some of are converted to molecular oxidants like hydrogen peroxide ($H_2O_2$), peroxynitrite (ONOO-), and hypochlorous acid (HOCl).

ROS are normally produced at a low level in all body cells through the electron-transport chain during aerobic respiration and by various constitutively active oxidases. Normally, they play an important physiological role at this low concentration, and are regulated by cellular antioxidants such as superoxide dismutase (SOD), catalase (CAT) for example. But sometimes it can happen an imbalance between ROS products and antioxidants in favor of ROS that can be due to either an overproduction of ROS or a lack of antioxidant or both. Such imbalance generally leads to molecules damaging (sugars, proteins, lipids, DNA) called oxidative stress which is responsible or implicated in many pathological conditions. Chronic oxidative stress has been considered as an indication in pathophysiological of diseases such as cancer, chronic inflammatory diseases, arthritis, neurodegenerative diseases, diabetes, and arthrosclerosis.

**TABLE 4.2** Examples of Redox Stimuli-Responsive Gene Delivery Systems

| Polymer Used | Type of System | Therapeutic Agent Used | References |
|---|---|---|---|
| (PEOz-PLA-g-PEI-SS) triblock copolymer-based: poly(2-ethyl-2-oxazoline)-poly(L-lactide) grafted with bioreducible polyethylenimine | Micelles | Minicircles DNA, doxorubicine | [43] |
| Disulfide-containing cross-linked polyethyleneimine (PEI-SS-CLs) | Polyplexes | pDNA | [50] |
| Hyaluronic acid-coated reducible hyperbranched poly-(amidoamine) (RHB) (HA/RHB/pDNA) | Nanoassembly | pDNA | [51] |
| Polyaspartamide-based disulfide-containing brushed polyethyleneimine (P(Asp-Az)X-SS-PEIs) | Polyplexes | pDNA | [52] |
| Arginine-conjugated poly(cystamine-bis-acrylamide–diamino hexane) (poly(CBA-DAH-R)) | Polyplexes | siRNA | [53] |
| Block co-polymer from PLGA-s-s-PEGMA (Poly(DL-lactide-co-glycolide)-s-s-poly(ethyleneglycolmethacrylate)) | Core-shell nanoparticle | pDNA | [54] |
| Disulfide bond bearing polyamido amines | Polyplex diameter of 100 to 250 nm | pDNA siRNA | [65, 66] |
| Covalent crosslinked poly(l-lysine) PLL/pDNA polyplexes | PLL/pDNA polyplexes | pDNA | [67] |
| DSP (3,3'-dithiobissuccinimidyl propionate) crosslinked PEI/ pDNA polyplexes | Polyplexes; liver-specific distribution and delivery | pDNA | [68] |

Reactive Oxygen Species (ROS) responsive delivery is a strategy based on the sensitivity of some moiety such as polysaccharides, thio-ether, thioketal, boronic acid (BA)/ester, l-methionine, and others for reactive oxygen species. When the responsive moieties are in contacts with an environment that contains oxygen species there is a reaction that leads to a change of the polymer properties, it can be a solubility switch or even a biodegradation of the polymer, which lead to the release of the loaded content of the polymers. Some systems based on this strategy have been developed in gene delivery as an attempt to overcome difficulties related to gene release and enhance the transfection efficiency.

Liu et al. have developed a charge-reversal concept based on a change of a cationic polymer to a fully negatively charged molecule in response to intracellular ROS, with the purpose to accelerate the release of the packed DNA and to minimize interference with the DNA transcrip-tion machine. The concept exploited a reactive oxygen species (ROS)-labile charge-reversal polymer, poly [(2-acryloyl) ethyl (p-boronic acid benzyl)diethylammonium bromide] (B-PDEAEA) which undergoes oxidation of the BA group, that led to the quaternary ammonium change into a tertiary amine, which catalyzed the ester group hydrolysis and produces poly(acrylic acid) (PAA). The ROS-responsive B-PDEAEA/ DNA polyplexes have shown high gene transfection efficiency than gold standard [55].

Thioketal nanoparticles (TKNs) have been designed for the localized delivery of siRNA orally administered to intestinal inflammatory sites. The nanoparticles have been developed from an ROS-responsive polymer, poly-(1,4-phenyleneacetone dimethylene thioketal) (PPADT) polymer, which is a thioketal containing polymer, with thioketal as the ROS sensitive moiety. The ROS-responsive thioketal linkage is stable in acidic, basic, and enzyme-catalyzed reactions, but is cleaved in an unusual elevated ROS environment and leads to DNA release. The TKNs loaded with siRNA have shown against the pro-inflammatory cytokine tumor necrosis factor-alpha (TNF-α) a diminution of TNF-α messenger RNA levels in the colon and protected mice from ulcerative colitis [56].

Yu et al. have developed an ROS ($H_2O_2$)-responsive micelles for the efficient delivery of therapeutics in cancer cells. The micelles was composed of methoxy polyethylene glycol-b-poly(diethyl sulfide) (mPEG-PS) copolymers, which facilitates endosomal escape of the therapeutics in cancer cells. The polymer was doted of the capacity to convert from hydrophobic to hydrophilic when in presence of high ROS levels such as in cancer cells. it destabilizes the cancer intracellular endosome membrane after cellular uptake [57].

## 4.2.1.5   ENZYME RESPONSIVE POLYMERS

Enzymes are proteins that play essentials roles in many biological processes such as metabolic process and many others under mild conditions, pH 5–8 and 37°C. They act like catalysts for numerous reactions in the living organism and they possess the properties to be selective and specific to a particular substrate. Thus, they can be used as trigger molecules for nucleic acids agents release at localized site, by incorporating their substrates in the polymeric carriers. Enzyme responsive carrier can be achieved by either graft/modify polymer with enzyme responsive moiety or by making the carrier with enzyme degradable polymers. As for the others bioresponsive system the polymeric materials used are mainly natural's polymers such as CS, pectin, dextran, cyclodextrin, and/or synthetic polymers such as PLL, PNIPPAAM, and PEG.

To achieve such delivery based enzyme responsiveness, various classes of enzymes are generally used such as hydrolases, reductases, proteases, endopeptidases, elastases, etc. some examples:

The matrix metalloproteinases (MMPs), which cleaves peptide bonds between non-terminal amino acids and are overexpressed in inflammation and cancer, will react with a peptide incorporated in the polymer (enzyme cleavable site) and will lead in the modification of the structure and the morphology of the polymer [72].

Hennink et al. have exploited MMPs responsiveness to prepare an *PEG-sheddable polyplex micelles* based on diblock copolymer $PEG_{227}$-GPLGVRG-PAsp $(DET)_{64}$, in which PAsp(DET) binds the pDNA to form core structure, which is shielded by PEG linked with the GPLGVRG peptides, which is the substrate-specific for MMPs [73, 74]

Chu, Johnson, & Pun [75] have prepared a polyplex containing pDNA based on N-(2-hydroxypropyl) methacrylamide (HPMA) polymer copolymerized with a cationic oligomer comprising cathepsin B-substrate peptide. They reported that the polyplex facilitated the release of pDNA in the presence of cathepsin B [75].

Zhu et al. have prepared a HMS/PLL particles which is a combination of hollow mesosporous silica and enzyme sensitive poly L-lysine polymer with the purpose to make a responsive carrier for simultaneous delivery of gene and therapeutic in cancer therapy. Those HMS/PLL particles react with alpha-chemotrypsin to release his loading [76].

### 4.2.1.6 ATP-RESPONSIVE POLYMERS

The adenosine triphosphate (ATP) is an important nucleotide molecule which consists of Adenosine and triphosphate. It is responsible for storage and transport of energy within cells, and also plays a role in nucleic acids synthesis.

The ATP molecule is the most abundant nucleotide in the body and is obtained by different processes such as oxidative phosphorylation in mitochondria, a reaction during which, one molecule of glucose gives two molecules of ATP which later are hydrolyzed in ADP (adenosine diphosphate). That hydrolysis reaction is responsible for the energy supply required to maintain metabolisms such as muscle contraction, active transport, and glycolysis.

The large gap in the ATP concentration between the intracellular and the extracellular medium is the main parameter that makes possible the design of ATP-responsive polymers for nucleic acid delivery.

Naito et al. [77], have developed an ATP-responsive siRNA micelle based on PEG-b-P(Lys/FPBAX)$_{42}$ complex in which siRNA have been conjugated with PEG-b-PLys initially modified in PLys chain with phenyl boronic acid (PBA). PBA as ATP responsive moiety because it is capable of forming reversible ester linkage with 1,2- or 1,3-cis-diols molecules like ATP and ribose molecules at the 3' end of siRNA. The micelles formed have showed a selective release of payloads siRNA at high concentration of ATP (<1mM). They have also reported that the PEG-b-P(Lys/FPBA$_{23}$)$_{42}$ complex have shown an encouraging dose-dependent silencing of the proto-oncogene polo-like kinase 1 (PLK-1) gene, in a human renal carcinoma cell line (OSRC-2) with minimal cytotoxicity [77].

### 4.2.1.7 OTHER STRATEGIES AND RECENT TRENDS IN BIORESPONSIVE POLYMERS BASED GENE DELIVERY

1. **Multi Stimuli-Responsive Systems:** This approach is based on the combination of more than one stimulus for the delivery of

nucleic acids, with the purpose to overcome intracellular multistep barriers that limit high levels of gene expression. It involves the modification of polymer with specific groups or moieties in the way that it can sense different stimuli at the same time, which can be pH, temperature, ROS, or another one. This approach could be the best one to achieve high transfection efficiencies similar to those of viruses.

Dual stimuli-responsive systems are the widely developed, many researchers have attempted to achieve this concept, and most of them have exploited the reducibility of disulfide bonds and the sensitivity of maleic amides groups to pH, to developed dual pH-sensitive and reducible responsive systems for the delivery of Nucleic Acids.

It is the case of Wagner et al. have developed a PEGylated polycation conjugated with melittin via disulfide bonds and with its membrane activity masked with maleic amides, for the delivery of siRNA Neuro2A-EGF plus cells [78].

Another approach to attempt dual responsiveness, involves the coating of reducible complexes with pH-responsive polymer [79] have reported the synthesis of virus-like particles which incorporates Reducible polyplexes made from peptidic RPCs, coated with pH-responsive polymers, bearing acetal linkers.

Many others multi-responsive systems exploiting different approach and sensing different stimuli have been developed:

A dual-triggered microenvironment-responsive delivery system that has been developed and demonstrated to respond to lowered pH (tumor extracellular pH) and MMP-2 [80].

Cho et al. have developed sequentially activatable polymeric gene complexes which is a quadruple stimuli-responsive system, responding to pH, light, thiol, and ROS. It is a highly efficient polymeric nanosystem (PhA@RPC/pDNA complex) which includes a photosensitizer (pheophorbide A (PhA))-loaded thiol-degradable polycation (high molecular weight (HMW)), bioreducible poly(ethylenimine (RPC)) and cytomegalovirus (CMV) promoter-equipped pDNA. That system could be promising tools for gene expression [81].

2.  **Adenovirus Complexes with Bioresponsive Polymers:** Nonviral mediated gene delivery using bioresponsive polymers have

shown promising results, in term of providing numerous design possibility, achieving high efficiency, good stability, less toxicity and many others advantages. However, these methods are currently less effective than the use of viral vectors even if provides a better safety. It has been thought that bioresponsive polymers integrated or associated to viruses would likely play an important role in cancer gene therapy, by combining both advantages of viruses and smart polymers such as conferring the virus increased tumor-specificity, reduced immunogenicity, effective reproducibility, and a simple quality control process.

Adenovirus which is the most efficient among viral vectors mediated gene delivery is also doted of oncolytic properties and have been mainly exploited for its innate cancer-specific killing potency. However, it faces some challenges such as immune response of organism which leads to reduce its efficiency, and also it causes cytotoxic reactions over hepatocytes some others sides effects. Thus, to overcome these problems, developed smart Adenovirus nanocomplexes have become a popular strategy, which mainly focus on surface modification (coating, conjugation) of the viral vectors by bioresponsive cationics polymers such as PEI, PLL which are the frequently used. Several studies have successfully exploited that method and showed some good results.

Moon et al. have developed a complex of adenovirus complexes with PPBCA polymer (Ad-PPCBA), to prevent immune response against adenovirus after intratumoral and systemic administration for the treatment of primary and metastatic cancer. The complex consisted of a pH sensitive and bioreducible polymer (mPEG-piperazine-N, N0-cystaminebisacrylamide) PPCBA, in which piperazine and N, N0 -cystaminebisacrylamide act respectively as pH sensitive and bioreducible moieties. The complex (Ad-PPCBA) was able to target low pH environment of cancer cells and also showed a lower toxicity and an enhanced therapeutic efficiency in both CAR-positive and negative cells (coxsackie and adenovirus receptor), which was significantly increased in the low pH environment of cancer cells [82].

In the same way, Choi et al. have developed the pH-responsive (Ad-PEG*b*PHF) nanocomplex, in which Adenovirus was coated with the pH-sensitive block copolymer, methoxy poly(ethylene glycol)-b-poly(L-histidine-co-L-24 phenylalanine) (PEG*b*PHF). Ad/PEG*b*PHF complexed with an oncolytic Ad expressing VEGF promoter-targeting transcriptional repressor (KOX) KOX/ PEGbPHF showed a higher antitumor activity and angiogenic effects at pH 6.4 than naked KOX and also when placed at pH 7.4. According to these results, they reported that Ad/PEG*b*PHF could be a safe and effective means for delivering hybrid therapeutic Ad vectors via systemic administration [83].

## ACKNOWLEDGMENT

I am grateful to Dr. Pinkal Patel and Mr. Nadim Chippa for their help to draw chemical structure in the book, and I would like to thank all the contributors to complete this chapter on time.

## KEYWORDS

- adenosine diphosphate
- charge conversion property
- cysteine
- Epstein Barr virus
- glutathione
- hepatitis B virus

## REFERENCES

1. Legendre, J. Y., Haensler, J., & Remy, J. S., (1966). *Nonviral Vectors of Gene Therapy, Medicine Science, 12,* 1331–1341.
2. Eliyahu, H., Barenholz, Y., & Domb, A. J., (2005). Polymers for DNA delivery. *Molecules, 10,* 31–64.
3. Schatzlein, A. G., (2003). Targeting of synthetic gene delivery systems. *Journal of Biomedicine and Biotechnology, 2,* 141–158.
4. Felgner, P. L., Gadek, T. R., Holm, M., Roman, R., Chan, H. W., Wenz, M., Northrop, J. P., Ringold G. M., & Danielsen, M., (1987). Lipofection: A highly efficient,

lipid-mediated DNA-transfection procedure. *Proceedings of the National Academy of Sciences of the United States of America, 84,* 7413–7417.

5. Stamatatos, L., Leventis, R., Zuckermann, M. J., & Silvius, J. R., (1988). Interactions of cationic lipid vesicles with negatively charged phospholipid vesicles and biological membranes. *Biochemistry, 27,* 3911–3925.

6. Alvarez-Lorenzo, C., & Concheiro, A., (2014). Smart drug delivery systems: From fundamentals to the clinic. *Chemical Communications, Journal of the Chemical Society, 50,* 7743–7765.

7. Behr, J. P., (1994). Gene transfer with synthetic cationic amphiphiles: Prospects for gene therapy. *Bioconjugate Chemistry, 5,* 381–389.

8. Boussif, O., Lezoualch, F., Zanta, M. A., Mergny, M. D., Scherman, D., Demeneix, B., & Behr, J. P., (1995). A versatile vector for gene and oligonucleotide transfer into cells in culture and *in vivo*: Polyethylenimine. *Proceedings of the National Academy of Sciences of the United States of America, 92,* 7297–7301.

9. Mitsuo, O., Yasuhiro, K., Yuichi, L., & Narumi, U., (2013). New vectors for gene delivery: Human and mouse artificial chromosomes. *Advancement in Genetic Engineering, 2,* 3.

10. Suh, J., Paik, H. J., & Hwang, B. K., (1994). Ionization of poly(ethylenimine) and poly(allylamine) at various pH's. *Bioorganic Chemistry, 22,* 318–327.

11. Guan, X., Guo, Z., Lin, L., Chen, J., Tian, H., & Chen, X., (2016). Ultrasensitive pH triggered charge/size dual-rebound gene delivery system. *Nano Letters, American Chemical Society, 16*(11), 6823–6831.

12. Takemoto, H., Miyata, K., Nishiyama, N., & Kataoka, K., (2014). Bioresponsive polymer-based nucleic acid carriers. *Advances in Genetics, 88,* 281–323.

13. Lee, Y., Miyata, K., Oba, M., Ishii, T., Fukushima, S., Han, M., Koyama, H., Nishiyama, N., & Kataoka, K., (2008). Charge-conversion ternary polyplex with endosome disruption moiety: A technique for efficient and safe gene delivery. *Angewandte Chemie International Edition, 47,* 5163.

14. Uchida, S., Itaka, K., Uchida, H., Hayakawa, K., Ogata, T., Ishii, T., et al., (2013). *In vivo* messenger RNA introduction into the central nervous system using polyplex nanomicelle. *PLoS One, 8*(2), e56220.

15. Ge, Z., Chen, Q., Osada, K., Liu, X., Tockary, T. A., Uchida, S., et al. (2014). Targeted gene delivery by polyplex micelles with crowded PEG palisade and cRGD moiety for systemic treatment of pancreatic tumors. *Biomaterials, 35,* 3416–3426.

16. Agarwal, S., Zhang, Y., Maji, S., & Greiner, A., (2012). PDMAEMA based gene delivery materials. *Materials Today, 15*(9), 381–393.

17. Samsonova, O., Pfeiffer, C., Hellmund, M., Merkel, O. M., & Kissel, T., (2011). Low molecular weight pDMAEMA-*block*-pHEMA block-copolymers synthesized via RAFT-polymerization: Potential non-viral gene delivery agents? *Polymers, 3*(2), 691–718.

18. Car, A., Baumann, P., Duskey, J. T., Chami, M., Bruns, N., & Meier W., (2014). pH-responsive PDMS-*b*-PDMAEMA micelles for intracellular anticancer drug delivery. *Biomacromolecules, 15*(9), 3235–3245.

19. Yu, H., & Wagner, E., (2009). Bioresponsive polymers for nonviral gene delivery. *Current Opinion in Molecular Therapeutics, 11*(2), 161–178.

20. Lin, S., Du, F., Wang, Y., Ji, S., Liang, D., Yu, L., & Li, Z., (2008). An acid-labile block copolymer of PDMAEMA and PEG as potential carrier for intelligent gene delivery systems. *Biomacromolecules, 9*(1), 101–115.

21. Maeda, Y., Pittella, F., Nomoto, T., Takemoto, H., Nishiyama, N., Miyata, K., et al. (2014). Fine-tuning of charge-conversion polymer structure for efficient endosomal escape of siRNA-loaded calcium phosphate hybrid micelles. *Macromolecular Rapid Communications, 35,* 1211–1215.

22. Rozema, D. B., Ekena, K., Lewis, D. L., Loomis, A. G., & Wolff, J. A. (2003). Endosomolysis by masking of a membrane-active agent (EMMA) for cytoplasmic release of macromolecules. *Bioconjugate Chemistry, 14,* 51–57.

23. Sanjoh, M., Hiki, S., Lee, Y., Oba, M., Miyata, K., Ishii, T., et al., (2010). pDNA/poly(L-lysine) polyplexes functionalized with a pH-sensitive charge-conversional poly(aspartamide) derivative for controlled gene delivery to human umbilical vein endothelial cells. *Macromolecular Rapid Communications, 31,* 1181–1186.

24. Takemoto, H., Miyata, K., Hattori, S., Ishii, T., Suma, T., Uchida, S., et al., (2013). Acidic pH responsive siRNA conjugate for reversible carrier stability and accelerated endosomal escape with reduced IFNα-associated immune response. *Angewandte Chemie International Edition, 52,* 6218–6221.

25. Lynn, D. M., & Langer, R., (2000). Degradable poly(β-amino esters), synthesis, characterization and self-assembly with plasmid DNA. *The Journal of the American Chemical Society, 122*(44), 10761–10768.

26. Zhong, Z., Song, Y., Engbersen, J. F., Lok, M. C., Hennink, W. E., & Feijen, J., (2005). A versatile family of degradable non-viral gene carriers based on hyperbranched poly(ester amine)s. *Journal of Controlled Release, 109*(1–3), 311–329.

27. Nam, H. Y., Nam, K., Hahn, H. J., Kim, B. H., Lim, H. J., Kim, H. J., Choi, J. S., & Park, J. S., (2009). Biodegradable PAMAM ester for enhanced transfection efficiency with low cytotoxicity. *Biomaterials, 30*(4), 661–673.

28. Kim, H. K., Thompson, D. H., Jang, H. S., Chung, Y. J., Van den, & Bossche, J., (2013). pH-responsive biodegradable assemblies containing tunable phenyl-substituted vinyl ethers for use as efficient gene delivery vehicles. *Applied Materials and Interfaces, 5,* 5648–5658.

29. Su, X., Fricke, J., Kavanagh, D. G., & Irvine, D. J., (2011). *In vitro* and *in vivo* mRNA delivery using lipid-enveloped pH-responsive polymer nanoparticles. *Molecular Pharmaceutics, 8,* 774–787.

30. Yu, H., Zou, Y., Wang, Y., Huang, X., Huang, G., Sumer, B. D., Boothman, D. A., & Gao, J., (2011). Overcoming endosomal barrier by amphotericin B-loaded dual pH-responsive PDMA-b-PDPA micelle plexes for siRNA delivery. *Nano, 5,* 9246–9255.

31. Benoit, D. S., Henry, S. M., Shubin, A. D., Hoffman, A. S., & Stayton, P. S., (2010). pH responsive polymeric siRNA carriers sensitize multidrug resistant ovarian cancer cells to doxorubicin via knockdown of polo-like kinase. *Molecular Pharmaceutics, 7,* 442–455.

32. Parhiz, H., Hashemi, M., Hatefi, A., Shier, W. T., Farzad, A. S., & Ramezani, M., (2013). Arginine-rich hydrophobic polyethylenimine: Potent agent with simple components for nucleic acid delivery. *International Journal of Biological Macromolecules, 60,* 18–27.

33. Zhang, L., Xu, T., & Lin, Z., (2006). Controlled release of ionic drug through the positively charged temperature-responsive membranes. *The Journal of Membrane Science, 281*, 491–499.

34. Mathew, A. P., Cho, K., Uthaman, S., Cho, C., & Park, I., (2017). Stimuli-regulated smart polymeric systems for gene therapy. *Polymers, 9,* 152.

35. Akimoto, J., Nakayama, M., Sakai, K., & Okano, T., (2009). Temperature-induced intracellular uptake of thermoresponsive polymeric micelles. *Biomacromolecules, 10,* 1331–1336.

36. Lavigne, M. D., Pennadam, S. S., Ellis, J., Alexander, C., & Gorecki, D. C., (2007). Enhanced gene expression through temperature profile-induced variations in molecular architecture of thermoresponsive polymer vectors. *The Journal of Gene Medicine, 9,* 41–54.

37. Hinrichs, W., Schuurmans-Nieuwenbroek, N., Van De Wetering, P., & Hennink, W., (1999). Thermosensitive polymers as carriers for DNA delivery. *Journal of Controlled Release, 60,* 249–259.

38. Kurisawa, M., Yokoyama, M., & Okano, T., (2000). Gene expression control by temperature with thermo-responsive polymeric gene carriers. *Journal of Controlled Release, 69,* 127–137.

39. Bouchemal, K., Aka-Any-Grah, A., Dereuddre-Bosquet, N., Martin, L., Lievin-Le-Moal, V., Le Grand, R., Nicolas, V., Gibellini, D., Lembo, D., & Pous, C., (2015). Thermosensitive and mucoadhesive pluronic-hydroxypropylmethylcellulose hydrogel containing the mini-cd4 m48u1 is a promising efficient barrier against HIV diffusion through macaque cervicovaginal mucus. *Antimicrobial Agents and Chemotherapy, 59,* 2215–2222.

40. Aguilar, M. R., Elvira, C., Gallardo, A., Vazquez, B., & Roman, J. S., (2007). Smart polymers and their applications as biomaterials. *Biomaterials, Topics in Tissue Engineering, 3,* 1–27.

41. Chao, L., & Bo Lou. (2012). Bioreducible cationic polymers for gene transfection, biomedicine, In: Chao, L., (ed.), *Tech.* (pp. 81–104). https://www.intechopen.com/books/biomedicine/bioreducible-cationic-polymers-for-gene-therapy (accessed on 16 January 2020). doi: 10.5772/38846.

42. Ullah, I., Muhammad, K., Akpanyung, M., Nejjari, A., Luis Neve, A., Guo, J., Feng, Y., & Shi, C., (2017). Bioreducible hydrolytically degradable and targeting polymers for gene delivery. *The Journal of Materials Chemistry, 5,* 3251–3276.

43. Gaspar, V. M., Baril, P., Costa, E. C., De Melo-Diogo, D., Foucher, F., Queiroz, J. A., et al., (2015). Bioreducible poly(2-ethyl-2-oxazoline)-PLA-PEI-SS triblock copolymer micelles for co-delivery of DNA minicircles and doxorubicin. *Journal of Controlled Release, 213,* 171–191.

44. Oupicky, D., Parker, A. L., & Seymour, L. W., (2002). Laterally stabilized complexes of DNA with linear reducible polycations: Strategy for triggered intracellular activation of DNA delivery vectors. *The Journal of the American Chemical Society, 124*(1), 1–9.

45. Lin, C., Zhong, Z. Y., Lok, M. C., Jiang, X., Hennink, W. E., Feijen, J., & Engbersen, J. F. J., (2007a). Random and block copolymers of bioreducible poly(amido Amine) s with high- and low-basicity amino groups: Study of DNA condensation and buffer capacity on gene transfection. *Journal of Controlled Release, 123*(1), 61–75.

46. Kim, H., Nam, K., Nam, J. P., Kim, H. S., Kim, Y. M., Joo, W. S., & Kim, S. W., (2015). VEGF therapeutic gene delivery using dendrimer type bio-reducible polymer into human mesenchymal stem cells (hMSCs). *Journal of Controlled Release, 220,* 221–228.

47. Kommareddy, S., & Amiji, M., (2005). Preparation and evaluation of thiol-modified gelatin nanoparticles for intracellular DNA delivery in response to glutathione. *Bioconjugate Chemistry, 16,* 1423–1432.

48. Kommareddy, S., & Amiji, M., (2007). Poly(ethylene glycol)–modified thiolated gelatin nanoparticles for glutathione-responsive intracellular DNA delivery. *Nanomedicine: Nanotechnology, Biology and Medicin*e, *3,* 32–42.

49. Rozema, D. B., Lewis, D. L., Wakefield, D. H., Wong, S. C., Klein, J. J., Roesch, P. L., et al., (2007). *Proceedings of the National Academy of Sciences of the United States of America, 104,* 12982.

50. Liu, J., Jiang, X., Xu, L., Wang, X., Hennink, W. E., & Zhuo, R., (2010). Novel reduction-responsive cross-linked polyethylenimine derivatives by click chemistry for nonviral gene delivery. *Bioconjugate Chemistry, 21,* 1827–1835.

51. Gu, J., Chen, X., Ren, X., Zhang, X., Fang, X., & Sha, X., (2016). Cd44-targeted hyaluronic acid-coated redox-responsive hyperbranched poly(amido amine)/plasmid DNA ternary nanoassemblies for efficient gene delivery. *Bioconjugate Chemistry, 27,* 1723–1736.

52. Zhang, G., Liu, J., Yang, Q., Zhuo, R., & Jiang, X., (2012). Disulfide-containing brushed polyethylenimine derivative synthesized by click chemistry for nonviral gene delivery. *Bioconjugate Chemistry, 23,* 1290–1299.

53. Kim, S. H., Jeong, J. H., Kim, T. I., Kim, S. W., & Bull, D. A., (2008). Vegf siRNA delivery system using arginine-grafted bioreducible poly(disulfide amine). *Molecular Pharmaceutics, 6,* 718–726.

54. Saeed, A. O., Magnusson, J. P., Moradi, E., Soliman, M., Wang, W., Stolnik, S., Thurecht, K. J., Howdle, S. M., & Alexander, C., (2011). Modular construction of multifunctional bioresponsive cell-targeted nanoparticles for gene delivery. *Bioconjugate Chemistry, 22,* 156–168.

55. Liu, X., Xiang, J., Zhu, D., Jiang, L., Zhou, Z., Tang, J., Liu, X., Huang, Y., & Shen, Y., (2016). Fusogenic reactive oxygen species triggered charge-reversal vector for effective gene delivery. *Advanced Materials, 28,* 1743–1752.

56. Wilson, D. S., Dalmasso, G., Wang, L., Sitaraman, S. V., Merlin, D., & Murthy, N., (2010). Orally delivered thioketal-nanoparticles loaded with tnfα-siRNA target inflammation and inhibit gene expression in the intestines. *Gastroenterology, 138,* 35–36.

57. Yu, L. Y., Su, G. M., Chen, C. K., Chiang, Y. T., & Lo, C. L., (2016). Specific cancer cytosolic drug delivery triggered by reactive oxygen species-responsive micelles. *Biomacromolecules, 17,* 3040–3047.

58. Yu, H., & Wagner, E., (2009). Bioresponsive polymers for nonviral gene delivery. *Current Opinion in Molecular Therapeutics, 11*(2), 161–178.

59. Meyer, M., Philipp, A., Oskuee, R., Schmidt, C., & Wagner, E., (2008). Breathing life into polycations: Functionalization with pH-responsive endosomolytic peptides and polyethylene glycol enables siRNA delivery. *The Journal of the American Chemical Society, 130*(11), 3271–3273.

60. Walker, G. F., Fella, C., Pelisek, J., Fahrmeir, J., Boeckle, S., Ogris, M., & Wagner, E., (2005). Toward synthetic viruses: Endosomal pH-triggered deshielding of targeted polyplexes greatly enhances gene transfer *in vitro* and *in vivo*, *Molecular Therapeutics, 11*(3), 411–425.

61. Akinc, A., Thomas, M., Klibanov, A. M., & Langer, R., (2005). Exploring polyethylenimine-mediated DNA transfection and the proton sponge hypothesis. *The Journal of Gene Medicine, 7*(5), 651–663.

62. Rozema, D. B., Lewis, D. L., Wakefield, D. H., Wong, S. C., Klein, J. J., Roesch, P. L., et al., (2007). Dynamic polyconjugates for targeted *in vivo* delivery of siRNA to hepatocytes. *Proceedings of the National Academy of Sciences, 104*(32), 12981–12987.

63. Russ, V., Elfberg, H., Thoma, C., Kloeckner, J., Ogris, M., & Wagner, E. Novel degradable oligoethylenimine acrylate ester-based pseudodendrimers for *in vitro* and *in vivo* gene transfer (2008). *Gene Therapeutics, 15*(1), 11–29.

64. Russ, V., Gunther, M., Halama, A., Ogris, M., & Wagner E., (2008). 90. Oligoethyleniminegrafted polypropylenimine dendrimers as degradable and biocompatible synthetic vectors for gene delivery. *Journal of Controlled Release, 132*(2), 131–140.

65. Oupicky, D., & Parker, A. L., (2002). Seymour, laterally stabilized complexes of DNA with linear reducible polycations: Strategy for triggered intracellular activation of DNA delivery vectors. *The Journal of the American Chemical Society, 124*(1), 1–9.

66. Lee, Y., Mo, H., Koo, H., Park, J. Y., Cho, M. Y., Jin, G. W., & Park J. S., (2007). Visualization of the degradation of a disulfide polymer, linear poly(ethylenimine sulfide), for gene delivery. *Bioconjugate Chemistry, 18*(1), 11–18.

67. Ward, C. M., Read, M. L., & Seymour L. W., (2001). Systemic circulation of poly(llysine)/DNA vectors is influenced by polycation molecular weight and type of DNA: Differential circulation in mice and rats and the implications for human gene therapy. *Blood, 97*(8), 2221–2229.

68. Neu, M., Germershaus, O., Mao, S., Voigt, K. H., Behe, M., & Kissel, T., (2007). Crosslinked nanocarriers based upon poly(ethylene imine) for systemic plasmid delivery: *In vitro* characterization and *in vivo* studies in mice. *Journal of Controlled Release, 118*(3), 371–380.

69. Oishi, M., Nagasaki, Y., Itaka, K., Nishiyama, N., & Kataoka, K., (2005). Lactosylated poly(ethylene glycol)-siRNA conjugate through acid-labile β-thiopropionate linkage to construct pH-sensitive polyion complex micelles achieving enhanced gene silencing in hepatoma cells. *The Journal of the American Chemical Society, 127*(6), 1621–1625.

70. Oishi, M., Nagatsugi, F., Sasaki, S., Nagasaki, Y., & Kataoka, K., (2005). Smart polyion complex micelles for targeted intracellular delivery of PEGylated antisense oligonucleotides containing acid-labile linkages. *Chembiochem., 6*(4), 711–725.

71. Meyer, M., Zintchenko, A., Ogris, M., & Wagner, E., (2007). A dimethylmaleic acid-melittin-polylysine conjugate with reduced toxicity, pH-triggered endosomolytic activity and enhanced gene transfer potential. *The Journal of Gene Medicine, 9*(9), 791–805.

72. Chien, M. P., Thompson, M. P., Lin, E. C., & Gianneschi, N. C., (2012). Fluorogenic enzyme-responsive micellar nanoparticles. *Chemical Science, 3,* 2690–2694.

73. De Graaf, A. J., Mastrobattista, E., Vermonden, T., Van Nostrum, C. F., Rijkers, D. T., Liskamp, R. M., & Hennink, W. E., (2012). Thermosensitive peptide-hybrid ABC block copolymers obtained by ATRP: Synthesis, self-assembly, and enzymatic degradation. *Macromolecules, 45,* 842–851.

74. Li, J., Ge, Z., & Liu, S., (2013). Peg-sheddable polyplex micelles as smart gene carriers based on MMP-cleavable peptide-linked block copolymers. *Chemical Communications, 49,* 6974–6976.

75. Chu, D. S. H., Johnson, R. N., & Pun, S. H., (2012). Cathepsin B-sensitive polymers for compartment-specific degradation and nucleic acid release. *Journal of Controlled Release, 157,* 445–454.

76. Zhu, Y., Meng, W., Gao, H., & Hanagata, N., (2011). Hollow mesoporous silica/poly(L-lysine) particles for co-delivery of drug and gene with enzyme-triggered release property. *The Journal of Physical Chemistry C., 115,* 13630–13636.

77. Naito, M., Ishii, T., Matsumoto, A., Miyata, K., Miyahara, Y., & Kataoka, K., (2012). A phenylboronate-functionalized polyion complex micelle for ATP-triggered release of siRNA, *Angewandte Chemie International Edition, 51,* 10751–10755.

78. Meyer, M., Philipp, A., Oskuee, R., Schmidt, C., & Wagner, E., (2008). Breathing life into polycations: Functionalization with pH-responsive endosomolytic peptides and polyethylene glycol enables siRNA delivery. *The Journal of the American Chemical Society, 130,* 3272.

79. Soliman, M., Nasanit, R., Abulateefeh, S. R., Allen, S., Davies, M. C., Briggs, S. S., et al., (2012). Multicomponent synthetic polymers with viral-mimetic chemistry for nucleic acid delivery. *Molecular Pharmaceutics, 9*(1), 1–13.

80. Huang, S., Shao, K., Liu, Y., Kuang, Y., Li, J., An, S., Guo, Y., Ma, H., & Jiang, C., (2013). Tumor-targeting and microenvironment-responsive smart nanoparticles for combination therapy of antiangiogenesis and apoptosis. *Nano, 7,* 2860–2871.

81. Cho, H., Cho, Y., Kang, S., Kwak, M., Huh, K., Bae, Y., & Kang, H., (2017). Tempo-spatial activation of sequential quadruple stimuli for high gene expression of polymeric gene nanocomplexes. *Molecular Pharmaceutics, 14,* 842–855.

82. Choi, J. W., Kasala, D., Moon, C. Y., Jung, S. J., Kim, S., & Yun, C. O., (2015). Tumor microenvironment-targeting hybrid vector system utilizing oncolytic adenovirus complexed with pH-sensitive and bioreducible polymer. *Biomaterials, 41,* 51–68.

83. Choi, J. W., Joung, S. J., Kasala, D., Kyu Hwang, J., Hu, J. H., Bae, Y., & Yun, C. O., (2015). pH-sensitive oncolytic adenovirus hybrid targeting acidic tumor microenvironment and angiogenesis. *Journal of Controlled Release, 205,* 131–43, doi: 10.1016/j.jconrel.2015.01.005.

# CHAPTER 5

# Recent Developments in Bioresponsive Drug Delivery Systems

DRASHTI PATHAK and DEEPA H. PATEL

*Department of Pharmaceutics, Parul Institute of Pharmacy and Research, Faculty of Pharmacy, Parul University, P.O. Limda, Ta. Waghodia, Vadodara–391760, Gujarat, India, Phone: 02668-260287, Fax: 02668-260201, E-mails: deepaben.patel@paruluniversity.ac.in, Pateldeepa18@yahoo.com (D. H. Patel)*

## 5.1 BIO-RESPONSIVE DRUG DELIVERY IN CANCER THERAPEUTICS

Cancer is defined as a collection of life-threatening diseases, which are characterized by the uncontrolled division of abnormal cells (National Cancer Institute in 2005). It currently affects one in three British people, and one in four people die from it. While mortality associated with other major diseases, such as heart disease and infections, has improved since 1950, cancer mortality has shown little such progress and became the most common cause of death in Britain since 1969 in women, and 1995 in men (Office for National Statistics in 2004).

Therefore, nowadays dextrin-phospholipase A2 (PLA2) conjugates are used to investigate the potential of a polymer-PLA2 conjugate, either as an anti-cancer agent or for use as a trigger to promote the release of the liposome-encapsulated drug, in the context of Polymer Enzyme Liposome Therapy (PELT).

The search for better cancer treatments is ongoing and research largely falls into two categories; the design of novel tumor-specific molecular targets using genomics and proteomics research and improving drug delivery and tumor targeting. One approach currently showing promise in the fight against cancer these days is 'nanomedicine,' defined by the

European Science Foundation (ESF) as *the science and technology of diagnosing, treating, and preventing disease and traumatic injury, of relieving pain, and of preserving and improving human health, using molecular tools and molecular knowledge of the human body.* Nanomedicines have also been defined by the ESF as 'nanometer size scale complex systems, consisting of at least two components, one of which being the active ingredient.'

Polymer therapeutics are nanomedicines that are designed to deliver drug(s) specifically to tumors via the enhanced permeability and retention (EPR) effect, while reducing the exposure of normal tissue, and consequently reducing systemic toxicity, and prolonging drug half-life (reviewed in Duncan [31]). The term 'polymer therapeutic' was coined by Duncan and describes a class of therapeutic agents comprising of at least two components, including a water-soluble polymer covalently attached to an active constituent, such as a drug, protein, gene or peptide (reviewed in [31]).

Over the past decade, several polymer-protein conjugates have been in routine clinical use, while 11 polymer-drug conjugates have entered clinical development. Of these, N-(2-hydroxypropyl)methacrylamide (HPMA) copolymer-doxorubicin (DOX) has shown clinical activity in breast cancer and HPMA copolymer conjugates containing a combination of chemotherapy and aromatase inhibitors are already showing exciting synergistic activity.

## 5.1.1   NANOMEDICINES AND EPR EFFECT

The rationale for attaching a drug or protein to a polymer is that conjugation can reduce toxicity, it extends the plasma half-life, reduce immunogenicity and mask protein charge, and can also allow passive targeting of the entity to tumors by the EPR effect. Conjugation of a drug or protein to a polymer alters its pharmacokinetic profile, which can be exploited to promote passive tumor targeting. This arises as a result of greater vascular permeability of the tumor's blood vessels, which favors the extravasation of drugs. Additionally, tumors tend to have an extensive and disorganized vasculature, compared to normal tissues. The pathophysiology responsible for this increased permeability can be attributed to an increase in fenestrae, a discontinuous basement membrane, enlarged inter-endothelial junctions and an increased number of vesicles. Furthermore, the lymphatic drainage of the tumor is reduced, resulting in the pooling of circulating polymer-protein conjugates in the tumor interstitium.

Thus, the EPR effect can be exploited in the delivery of polymer therapeutics. While conventional, low molecular weight drugs can pass readily through the bloodstream, into the tissue or to be excreted, polymer therapeutics can accumulate in tumor tissues due to the EPR.

Ringsdorf also suggested that a targeting moiety could be incorporated into the polymer to enhance uptake by receptor-mediated endocytosis. This was shown by the polymer therapeutic PK2, which is composed of HPMA copolymer-Gly-Phe-Leu-Gly-DOX incorporating galactosamine. The galactosamine residue enhances uptake of the conjugate by the liver by targeting the hepatocyte asialoglycoprotein (ASGP) receptor, thus it was investigated as a treatment for liver cancer [1].

## 5.2 BIORESPONSIVE MATERIALS: ENGINEERING PARTICULATE MOIETIES

Pioneering studies using engineered materials for dosage and for spatially and/or temporally controlled drug release were carried out in the 1970s [2–4]. More recently, there has been a growing interest in the development of stimuli-responsive, 'smart' materials for a range of biomedical applications, including drug delivery, diagnostics, tissue engineering, and biomedical devices [5–7].

In bio-responsive drug delivery, the treatment efficacy of therapeutics is directly related to the administration method [8–11] which requires the development of advanced materials to achieve precision drug release (BOX 1). The improvement of medical diagnostics demands non-invasive or minimally invasive approaches based on stimuli-responsive materials that allow for real-time monitoring. The growing desire for tissue engineering and regenerative medicine leads to urgent needs for matrices endowed with the ability to communicate and interact with cells. While in the development of medical devices, the incorporation of high-performance biocompatible materials is crucial for improving the antibacterial and anti-inflammation capability and preventing the formation of biofilms.

In general, bioresponsive polymers can be deconstructed into functional motifs with biological sensitivities, which can be built into the desired formulations, scaffolds, or devices in a controlled manner using appropriate fabrication methodologies.

In the past decades, a considerable number of engineered biological particulates have been developed for targeted therapy, with some of them successfully entering clinical practice [12].

For mucosal-targeted delivery, recombinant *Lactobacillus acidophilus* was engineered to express the protective antigen of Bacillus anthracis, which was further fused with dendritic-cell-targeting peptide to recognize and bind to mucosal dendritic cells (DCs) [13]. Additionally, it has been revealed that certain strains of bacteria specifically colonize tumor cells, exhibiting natural tumor-targetability. Making use of their natural tumor tropism, these bacteria have been genetically modified to express protein therapeutics for cancer treatment [14]. Recently, researchers programmed a bacteria-based circuit to lyse and release anticancer toxins at the tumor site upon achieving the threshold population [15]. Similar to bacteria, viruses have also evolved into natural carriers with both specificity and efficiency. Cell-targeted carriers constructed from engineered viruses mainly take advantage of their natural tropism with a range of targets. For example, virus-like particles are particles self-assembled from virus-derived capsid or envelope proteins, whereas virosomes are spherical virion-like lipid bilayer vesicles containing virus-derived surface glycoproteins [13]. These vesicles have inherited the capability to specifically recognize and interact with target cells from their parental viruses.

Despite their potency, up to now the clinical use of these pathogen-based carriers has been limited to vaccine delivery. The potential immunogenicity has hindered their clinical translation [16]. Stem cells also possess tumor tropism and thus have been genetically engineered to express anticancer proteins specifically at tumor sites [17]. Owing to the capability of cells to endocytose nanoparticles or to absorb nanoparticles onto their surfaces, it is also possible to use stem cells as tumor-targeting carriers for delivering nanoparticles [18].

An important feature of cancer cells is homotypic tumor cell adhesion. In light of this, cancer cell membranes were coated onto the surface of PLGA nanoparticles for source-cell-targeted delivery [19]. Natural RNA transport vesicles, such as exosomes, have been genetically modified with targeting peptides for gene delivery [20]. Red blood cells (RBCs) are promising drug-delivery candidates that possess several appealing features, such as their long circulation time. In addition, their unique transporting characteristics enable RBC based delivery targeting RBC-eliminating cells, such as the reticuloendothelial system (RES) macrophages. Apart from directly using intact RBCs, RBC membranes have also been coated on PLGA cores for different applications, such as toxin absorption [21].

Moreover, synthetic nanoparticles have been coated with platelet membranes to target MRSA252 bacteria [22] and circulating tumor cells [23, 24]. As an essential component of the immune system, macrophages also display an inherent tendency to harbor diseased tissues attributed to the cellular secretion of signaling molecules, such as cytokines, and thus are natural vehicles for targeted drug delivery. Similar to nanoparticle-containing stem cells, nanoparticles can also be loaded into macrophages by means of phagocytosis [25].

The interaction between immune cells and target cells has been extensively explored for constructing bioresponsive vehicles. For example, the interactions among lymphocytes of the immune system have been used for developing a whole-cell sensing system [26]. In a recently developed method, live T cells expressing lymph-node-homing receptors were used as active carriers for delivering chemotherapeutic-loaded nanoparticles to cancerous lymphoid tissues [27]. T-cell therapy suffers from relatively low *in vivo* persistence and rapid function decline, and thus often requires co-administration of adjuvant drugs [28]. Using this strategy, adjuvant-loaded nanoparticles were conjugated onto the surface of T cells to enhance the efficacy of T-cell therapy [29].

By contrast, T-cell specificity could be effectively redirected through the genetically engineered expression of chimeric antigen receptors (CARs) on the surface of T cells. In CAR-T-cell therapy, the engineered CAR T cells are cultured and expanded in the laboratory and then infused into the patient after the desired population has been achieved.

## 5.3 BIO-RESPONSIVE POLYMERS: BIOMEDICAL APPLICATIONS

Use of polymers in biomedical materials applications—e.g., as prostheses, medical devices, contact lenses, dental materials, and pharmaceutical excipients—is long-established, but polymer-based medicines have only recently entered routine clinical practice [30–33]. Much innovative polymer-based therapeutics once dismissed as interesting but impractical scientific curiosities have now shown that they can satisfy the stringent requirements of industrial development and regulatory authority approval.

The term polymer therapeutics has been adopted to encompass several families of construct, all using water-soluble polymers as components for design: polymeric drugs [30], polymer-drug conjugates [34] polymer-protein conjugates [35] polymeric micelles to which drugs are covalently bound [36], and multicomponent polyplexes being developed as nonviral gene vectors. Nanosized medicines are more like new chemical entities than conventional drug delivery systems or formulations, which simply entrap, solubilize, or control drug release without resorting to chemical conjugation.

### 5.3.1   POLYMERS AND GENE THERAPIES

With the growing appreciation of the molecular basis of disease in the late 1980s, hopes for gene therapy began to gain momentum. Although viral vectors are still preferred for gene delivery, there has been a continuing hope that polymeric nonviral vectors can become a credible alternative: biomimetics delivering DNA safely without the threat of toxicity. Pioneering early research used simple polycationic vectors such as poly(L-lysine) and poly(ethyleneimine). A wide range of complex multicomponent, polymer-based vectors have been designed as gene delivery systems [37].

### 5.3.2   BIORESPONSIVE POLYMERS AND MOLECULAR IMAGING

Molecular imaging is a powerful tool to visualize and characterize biological processes at the cellular and molecular level *in vivo*. Molecular imaging involves the noninvasive visualization and quantitative detection of biomolecules *in vivo* by means of target-specific probes [38–41]. Valuable applications of molecular imaging are accurate disease detection, phenotyping, and staging by gathering information on molecular pathways underlying biological and cellular processes in the diseased tissue. Hence, molecular imaging is likely to play a pivotal role in the stratification of patients for personalized treatment. Furthermore, molecular imaging is clinically relevant in the discovery and development of drugs and for the real-time assessment of the efficacy of drug therapy [42, 43]. Further, it can contribute to improved interventions by image-guided drug delivery and image-guided surgery [44–47].

Finally, molecular imaging has an impact on the development of regenerative medicine and stem cell therapies. In most molecular imaging approaches, target-specific molecular probes are engineered to enhance image contrast at the target site. These conventional targeting molecular imaging probes consist of a ligand that binds to an endogenous molecular target and an imaging label for readout.

Typically, the molecular imaging probes are designed so that they can be injected intravenously for targeting biomolecules accessible via the blood circulation. For example, the active site of enzymes has been targeted with conventional probes, providing readout for enzyme abundance [48].

However, these probes lack the possibility of detecting enzymatic activity *in vivo* as they are typically limited to a 1:1 probe/target binding fashion. In that respect, a new subset of protease activatable molecular imaging probes has generated considerable attention. These probes have been directed to the active site of the enzymes resulting in the cleavage and activation of the imaging probe [49].

Importantly, this will lead to signal amplification since the molecular target can continuously activate the imaging probe and, moreover, it offers readout of enzyme activity instead of enzyme abundance.

### 5.3.3   OPTICAL PROBES ACTIVATED BY REACTIVE OXYGEN SPECIES

Reactive oxygen species play a crucial role in maintaining normal physiology. However, excessive production of ROS/RNS has been associated with a variety of pathological diseases, such as cancer and cardiovascular disease [50, 51]. Activatable fluorescent imaging probes for ROS detection are mainly based on autoquenched probes in which a fluorophore is coupled to a ROS trigger, resulting in a fluorescent molecule [52].

The reaction of these probes with ROS results in the activation of the fluorescent moiety. This approach has been extensively exploited, resulting in activatable fluorescent probes for a range of ROS, including hydrogen peroxide ($H_2O_2$) [53–60], peroxynitrite (ONOO-), superoxide ($O_2$), nitric oxide (NO), hypochlorous acid (HOCl) and hydroxyl radical (OH).

Other imaging approaches have focused on the colorimetric detection of ROS or used genetically encoded biosensors. ROS imaging methods have been applied in cell-culture experiments *in vitro*, but strategies to visualize ROS *in vivo* have been limited [61–64].

### 5.3.4  OPTICAL PROBES ACTIVATED UPON TARGET BINDING

The final class of bioresponsive fluorescent imaging probes shows a change in between a donor fluorophore and acceptor fluorophore or quencher upon target binding. For instance, various genetically encoded sensors based on mutants undergo a conformational change upon target binding and allowed the *in vitro* ratiometric detection of, for example, ATP, glucose, and metal ions like calcium, zinc, and cadmium.

Another strategy is focused on oligonucleic acid-based aptamer probes in which a fluorescent label is efficiently activated upon a conformational change of the probe induced by binding to a biomarker.

Alternatively, micellar, and liposomal formulations containing high concentrations of self-quenched fluorescent dyes have been developed that destabilize and release the fluorescent dyes upon target binding, cellular internalization, and subsequent nanocarrier degradation. Similarly, various probes have been developed that upon target binding and subsequent cellular internalization showed activation of the optical imaging label [65–76].

## 5.4  BIORESPONSIVE HYDROGELS AS EMERGING THERAPY

Biologically responsive (bioresponsive) hydrogels are stimuli-responsive hydrogels wherein the response of the hydrogel is triggered by the recognition of a biological agent by an immobilized biorecognition species. Biological agents may be biomolecules like glucose, large macromolecules like chymotrypsin, or even whole cells, such as vascular endothelial cells. Biorecognition species can be native or synthetic biomacromolecules such as enzymes, antibodies, nucleic acids, or small molecules such as metabolites or peptides.

The triggers may result in responses which include, among others: swelling or collapsing, degradation, mechanical deformation, optical density changes, or electrokinetic variations.

The response may be binary, as in the presence or absence of the biological agent at a particular threshold limit, or it may be scalable with the chemical potential or activity of the biological agent. Bioresponsive hydrogels link hydrogel responses to the innate specificity of bio-recognition reactions which enables them to elicit a response upon meeting specific criteria. This creates a "smart" material because it is able to sense surrounding biological agents and can remain in stasis or become dynamic depending on changes in the surrounding environment.

The engineering of bioresponsive hydrogels is often difficult because: (i) molecular and architectural design must be properly specified and fabricated, (ii) there are biologically imposed constraints in conditions of materials processing, and (iii) there are complex interactions between the hydrogel matrix, the conferring bioactive species, the targeted biological agent, and the solvent (usually water) during performance.

Successful design requires multiple length scale theoretical and experimental modeling to control kinetic, thermodynamics, and transport phenomena. Additionally, immobilization of the biorecognition species must be achieved while maintaining its bioactivity. The utility of bioresponsive hydrogels is vividly apparent in their applications. These hydrogels have been used in tissue engineering in mediated drug delivery and biosensing applications.

In tissue engineering, bioresponsive hydrogels have been used as biomimetic matrices, which promote cellular proliferation but also biodegrade over time. In drug delivery, it is commonly desired to deliver a drug in specified doses to a specific site, organ, or biological system [77–93].

Bioresponsive hydrogels are capable of location-specific delivery because they are able to release the therapeutic drug payload when they are triggered by biologically derived, site-specific stimuli. In sensor applications, bioresponsive hydrogels can produce a measurable response when a specific analyte of biological origin is present.

## 5.4.1   THE NEXT GENERATION HYDROGELS

The historical development of bioresponsive hydrogels has proceeded with incremental advances to a broadening base of scientific knowledge. Responsive hydrogels began with polymers which were able to sense and respond to environmental stimuli such as temperature and pH (Generation 1).

Further development led to bioresponsive hydrogels which sense biological agents pertinent to specific biological systems, pathways, or processes (Generation 2). Bioresponsive hydrogels have improved and forged new applications in a variety of fields; however, to continue their advancement a fundamental design concept is required in the information flow scheme. That is, the inclusion of closed control loops integrated into the hydrogel such that the hydrogel material and the surrounding environment operate as a cohesive system (Generation 3).

Gen3 hydrogels' responses effect and are affected by upstream or downstream processes in a manner which potentiates or attenuates biological activity accordingly. These materials will function as a synthetic metabolic pathway capable of therapeutically administering drugs personalized to a patient's biological activity. The conceptualization of an integrated feedback loop has existed for some time, and proposed platforms based on pH sensitivity or binding affinity have been demonstrated experimentally.

In all, these examples have been building towards full biomolecular integration; however, *in vitro* and *in vivo* demonstration of full integration is only beginning to be developed. This is exemplified in the glucose-responsive hydrogels of Guiseppi-Elie et al. where insulin was released based on exposure to glucose emulating the natural endocrine response. However, because the necessary metabolic pathways needed to affect the solution concentration of glucose were not present, the system's response (a change in the rate of release of insulin) was not informed by a change in the concentration of glucose [94–97].

### 5.4.2 CONTROLLED CLOSED LOOP THEORY FOR BIORESPONSIVE HYDROGELS

Control theory in Gen3 hydrogels is applied to two processes, the intrinsic metabolic process and the engineered sense and response of the hydrogel. Control theory in metabolic processes is a continually growing field with developed models for natural metabolic feedback and feed-forward control loops.

Control theories are being used to understand processes such as natural, closed-loop blood glucose control, neurological disease, and regulation in cellular networks. Conversely, the control theory around Gen3 hydrogels has not been extensively developed. The "programming" of Gen3 hydrogels is fundamentally different as tuning of physical parameters, such as diffusion constants, loading concentrations or swelling properties, will be used to effect time constants, $\tau_i$, and gains, $K_i$ [98–100].

### 5.4.3 BIOSPECIFICITY CONFERRING

The design of bioresponsive hydrogels seeks to unite the conferred capacity for biorecognition of a bioreceptor with the capacity for a tailored response using unique macromolecular structures and architectures of the hydrogels. Biorecognition may be conferred through the immobilization (physical entrapment, polyelectrolyte complex formation, covalent bonding, or adsorption) of biologically active moieties such as peptides, nucleic acids, enzymes, antibodies, antibody fragments, sub-cellular fragments, or whole cells. Most commonly, these are molecular entities such as peptides and nucleic acids which may be pendant to the hydrogel network backbone or may serve as covalent or virtual crosslinks. Hennink and Nostrum have reviewed cross-linking methodologies in hydrogels.

A widely used application of pendant peptide sequences is the attachment of the peptide Arg-Gly-Asp (RGD) which promotes receptor-mediated cellular attachment via integrins.

Additionally, peptide sequences can be immobilized within the hydrogel to create protease degradable hydrogels.

A common example is the immobilization of a bioactive peptide that may be cleaved by an external proteolytic enzyme thus releasing a drug that diffuses out of the hydrogel.

The foregoing work is now being extended to the use of biologically responsive "smart" hydrogels, wherein, negative feedback control is achieved by enzyme-mediated release of a factor which regulates the activity of a biological species outside the hydrogel, subsequently reducing the release rate of the factor. Living systems possess molecular mechanisms which allow them to maintain optimized physiological conditions, despite variations and disturbances to their environment. Homeostasis is thus a self-adjusting mechanism involving negative feedback control.

The design, synthesis, and fabrication of biologically responsive hydrogels to achieve homeostasis in external biological activity is a novel, and until now, an unexplored concept. A reverse of the foregoing, the immobilization of an enzyme within the hydrogel, is likewise feasible. Such configurations are commonplace in the construction of enzyme amperometric biosensors. In this case, the diffusible substrate enters the hydrogel, reacts, and generates a product that causes a change within the hydrogel.

In the exampled work of Guiseppi-Elie et al. glucose oxidase (GOx) was co-immobilized by physical entrapment along with insulin within a pH-responsive p(HEMA-co-DMAEMA) hydrogel. The mesh size was designed, in the un-protonated state, to permit ready diffusion of glucose into the hydrogel with little diffusion of insulin out. The catalytic action of GOx acting upon glucose results in the generation of gluconic acid leading to protonation of the tertiary amine of DMAEMA and further swelling of the hydrogel.

The expanded hydrogel, through an enhanced diffusion coefficient, then released insulin. The final two immobilization techniques for bioresponsive hydrogels are binding and adsorption. In binding, one member of the binding pair, such as biotin or protein-G, is first covalently immobilized and serves to immobilize the recognition molecule such as streptavidin. These species serve as cross-linking agents and by competitive binding can swell or collapse the hydrogel [101, 102].

Adsorption immobilization of biomacromolecules has been thoroughly investigated, and it has been found that important factors controlling adsorption and desorption are hydrophilic/hydrophobic and electrostatic interactions.

## 5.5 ELECTROCHEMICAL RESPONSE

Due to high biocompatibility, hydrogels have been widely used in biosensors, both as an interface between tissue and electrode and also as a host for bioactive molecules [134]. By incorporating oxidoreductase enzymes into the polymer matrix, the biospecificity of the enzyme is conferred to the transducer, and the enzyme is stabilized in a hydrated 3-D milieu.

In amperometric biosensors, the electrodes act as an eventual source or sink for the electrons being transferred by the enzyme. Biosensors which measure analyte concentrations have been designed based on other electrical properties such as potential or impedance.

To increase sensitivity, reduce interfacial impedance, reduce interference from endogenous and exogenous species, and improve biocompatibility, electroconductive polymers such as polyaniline, polythiophene, and polypyrrole, have been synthesized into the voids of the hydrogel matrices.

The resulting electroconductive polymer hydrogel composites also show the potential for programmed release of loaded drugs under modest voltage stimulation. This is particularly useful in the development of prosthetic electrodes for deep brain stimulation, intraocular implants, and cochlear implants. Another approach to facilitate electron transfer in biosensors is the conjugation of the redox enzyme to *single-walled carbon nanotubes* (SWNT). Enzyme-SWNT conjugates have been shown to increase the peak current; additionally, graphene-based biosensors are also being developed based on similar principles [103, 104] and maybe incorporated into hydrogels. These composite materials add new response modes to the bioresponsive hydrogels.

## 5.5.1 TARGETING HOMEOSTASIS

Hydrogels are polymers synthesized from highly hydrophilic monomers and/or pre-polymers into networks that imbibe large amounts of water, are passively biocompatible due to their favorable interaction with extra-cellular matrix (ECM) proteins, and are highly versatile in their range of possible biotechnical applications that exploit their swelling and deswelling dynamics [103].

Because of these characteristics, hydrogels have been extensively studied for a wide range of biomedical applications that include controlled drug delivery [97–99], tissue engineering, and regenerative medicine [98] and diagnostic biomedical biosensors.

Hydrogels have also evolved over the years through several generations wherein increased conferred functionality, control of polymer architecture and improved processing has enabled new and innovative applications resulting in the moniker of "smart materials." Among these smart applications is one wherein the ability to sense and react with a biological entity is integrated with the physicochemical responses of the hydrogel.

Biologically responsive, or bioresponsive, hydrogels operate by first sensing the chemical activity of a biological species in its environment, through catalysis or binding, and then couples the biomolecular recognition to transduction of that information with a response that seeks to restore equilibrium or a pseudo-steady state within the hydrogel. In this sense, they are Generation 2 drug release platforms that supersede Generation 1 drug release platforms, which rely on loading and passive release of a drug from an engineered microform. Of particular interest is the use of micro- and nano-hydrogels to sense and respond to spatiotemporal perturbations of biological activity as this enables the development of therapeutics and diagnostics at molecular and cellular length scales. Such systems are well described and are actively pursued in the literature.

The development of drug delivery systems that are based on self-regulation and that are made responsive to an externally derived or environmentally-based biological activity that dictates the release rate of the drug has long been a goal of drug delivery designers. This is particularly relevant in glucose-responsive insulin delivery, wherein the rate of release of insulin may be governed by the bathing concentration of glucose, analogous to an all synthetic artificial pancreas.

Recently, an *in vivo* demonstration of this objective that uses the glucose (sense)-insulin (deliver) system was reported; however, engineering control of key delivery parameters of such systems is unexplored.

Generation 3 drug delivery systems go beyond environmental control of the release rate to focus attention on achieving physiological stasis of the biological species that is the target of drug action. Self-regulated delivery systems imply a nascent ability to change the release profile pursuant to the amount of or extent of the release. Generation 3 systems use the released drug to modulate the activity of the actuating biological factor. In this sense, it is distinguishable from self-regulating systems whereupon the release kinetics is controlled by the environmental conditions.

The Generation 3 drug delivery system actually controls the environmental conditions and seeks to achieve homeostasis, that is, it targets the achievement of a particular activity of the actuating biological species within the environment that had elicited the release response in the first place. This requires the implementation of feedback control. The information flow from sensing to response in Generation 2 bioresponsive hydrogels often involves dynamical concomitant perturbations such as chemical reactions or binding events; diffusive, convective, and/or migratory transport of species; mechanical deformation; and optical density changes.

Due to the high interdependency of such changes within bioresponsive hydrogels, explicit solutions derived from constitutive mathematical relations are invariably difficult and more often unresolved. Thus, the approach to engineering and optimization of bioresponsive hydrogels is commonly *in vitro* or *in vivo* experimentation coupled with mathematical simplifications for analysis. This approach, while effective, can be time-intensive and may result in suboptimal systems. Fortunately, the exponential increase in computational power over the past decades is enabling numerical resolutions to constitutive relations, which were previously inaccessible.

The coupling of finite element modeling (FEM) with bioresponsive hydrogels to investigate previously difficult or inaccessible relations has and will continue to be a robust and fruitful area of research. The complexity of the models has scaled in parallel to increasing transistor density and has offered insights into these systems. Initially, due to limited computational power, only uni-physical models were explored, such as anisotropic drug transport in complex geometries.

However, complete multiphysics models are currently being developed in silico to inform the engineering of bioresponsive hydrogels. To this point, an in silico model of enzymatically mediated release from a bioresponsive hydrogel was developed and is presented.

The three-step process of release occurs by (1) diffusion of an enzyme to and into a hydrogel matrix, (2) enzymatic cleavage of a tethered peptide-prodrug conjugate that is a substrate-specific to the transported enzyme, and (3) subsequent diffusive release of the active drug into the local environment. *In vitro* variations of the enzyme-mediated release include the use of pendent or polymer-backbone integrated substrates, degradation or gelation of the hydrogel matrix and dynamic swelling or collapsing of the hydrogel matrix.

## 5.6 FUTURE OUTLOOKS

At present, the core business of the pharmaceutical industry is still low-molecular-weight drugs (natural product extracts and synthetic molecules) and pro-drugs, particularly those amenable to oral administration, which provides patient convenience. As a rule, macromolecular drugs—such as proteins, polymer therapeutics, and gene therapies—are not orally bioavailable. Coupled with their chemical complexity and perceived technological difficulties, which made them unattractive development candidates for most large pharmaceutical companies up until the end of the 20th century.

But the FDA approved more macromolecular drugs and drug delivery systems than small molecules as new medicines in 2002 and 2003, which suggests that the tide has turned. In the 21st century, the time is ripe to build on lessons learned over the past few decades, and with the increased efforts of polymer chemists working in multidisciplinary teams, this will surely lead to the design of improved second-generation polymer therapeutics.

The polymer community's interest in synthetic and supramolecular chemistry applied to biomedical applications has never been greater. This has in part been driven by a rise in interest toward using dendrimers and nanotubes for applications in drug delivery and the need for bioresponsive polymers that can be designed as three-dimensional scaffolds for tissue engineering. Innovative polymer synthesis is leading to many new materials, but although they provide exciting opportunities, they also present challenges for careful characterization of biological and physicochemical characterization. For clinical use, it is essential to identify biocompatible synthetic polymers that will not be harmful in relation to their route, dose, and frequency of administration.

For many years the general cytotoxicity, haematotoxicity, and immunogenicity (cellular and humoral) of water-soluble polymers have been widely studied. Before clinical studies, rigorous preclinical toxicity testing of each candidate has also been mandatory. However, it is becoming evident that synthetic polymers can display many subtle and selective effects on cells affecting a diverse range of biochemical processes. Such effects may be relatively weak so that they do not cause major toxicity. Studies are assessing the pharmacogenomic effects of polymers.

The development of analytical techniques that can accurately characterize polymer therapeutics in terms of identity, strength, stability, and structure in real-time (to allow correlation with biological properties) has proved a real challenge in itself. Atomic-force microscopy has already begun to demonstrate an ability to provide structural and physicochemical information for a range of synthetic and biopolymers.

## KEYWORDS

- **chimeric antigen receptors**
- **extracellular matrix**
- **finite element modeling**
- **hydroxyl radical**
- **peroxynitrite**
- **single-walled carbon nanotubes**

# REFERENCES

1. Elaine, L. F., (2013). *Bioresponsive Polymer-Phospholipase A2 Conjugates as Novel Anti-Cancer Agents* (pp. 1–30). Prouest LLC.
2. Langer, R., & Folkman, (1976). Journal of Polymers for the sustained release of proteins and other macromolecules, A pioneering study of the use of engineered materials for controlled drug delivery. *Nature, 263,* 797–800.
3. Yatvin, M., Weinstein, J., Dennis, W., & Blumenthal, (1978). Research design of liposomes for enhanced local release of drugs by hyperthermia. *Science, 202,* 1290–1293.
4. Brownlee, M., & Cerami, A., (1979). A glucose-controlled insulin delivery system: Semi synthetic insulin bound to lectin. *Science, 206,* 1190–1191.
5. Hoffman, A. S., (2013). Stimuli-responsive polymers: Biomedical applications and challenges for clinical translation. *Advanced Drug Delivery Review, 65,* 10–16.
6. Caldorera-Moore, M. E., Liechty, W. B., & Peppas, N. A., (2011). Responsive theranostic systems: Integration of diagnostic imaging agents and responsive controlled release drug delivery carriers. *Accounts of Chemical Research, 44,* 1061–1070.
7. Purcell, B. P., et al., (2014). Injectable and bioresponsive hydrogels for on-demand matrix metalloproteinase inhibition. *Nature Mater, 13,* 653–661.
8. Tibbitt, M. W., Dahlman, J. E., & Langer, R., (2016). Emerging frontiers in drug delivery. *Journal of American Chemistry Society, 138,* 704–717.
9. Kost, J., & Langer R., (2012). Responsive polymeric delivery systems. *Advanced Drug Delivery Reviews, 64,* 327–341.
10. Mitragotri, S., Burke, P. A., & Langer, R., (2014). Overcoming the challenges in administering biopharmaceuticals: Formulation and delivery strategies. *Nature Reviews Drug Discovery, 13,* 655–672.
11. Wang, S., Huang, P., & Chen, X., (2016). Hierarchical targeting strategy for enhanced tumor tissue accumulation/retention and cellular internalization. *Advanced Materials,* 7340–7364.
12. Yoo, J. W., Irvine, D. J., Discher, D. E., & Mitragotri, S., (2011). Bio-inspired, bioengineered and biomimetic drug delivery carriers. *Nature Reviews Drug Discovery, 10,* 521–535.
13. Mohamadzadeh, M., Duong, T., Sandwick, S. J., Hoover, T., & Klaenhammer, T. R., (2009). Dendritic cell targeting of *Bacillus anthracis* protective antigen expressed by *Lactobacillus acidophilus* protects mice from lethal challenge. *Proceedings of the National Academy of Sciences of the United States of America, 106,* 4331–4336.
14. Forbes, N. S., (2010). Engineering the perfect (bacterial) cancer therapy. *Nature Reviews Cancer, 10,* 785–794.
15. Din, M. O., et al., (2016). Synchronized cycles of bacterial lysis for *in vivo* delivery. *Nature, 536,* 81–85.
16. Iverson, N. M., et al., (2013). *In vivo* biosensing via tissuelocalizable near-infrared-fluorescent single-walled carbon nanotubes. *Nature Nanotechnology, 8,* 873–880.
17. Bago, J. R., et al., (2016). Therapeutically engineered induced neural stem cells are tumor-homing and inhibit progression of glioblastoma. *Nature Communications, 7,* 10593.

18. Roger, M., et al., (2010). Mesenchymal stem cells as cellular vehicles for delivery of nanoparticles to brain tumors. *Biomaterial, 31*, 8393–8401.

19. Fang, R. H., et al., (2014). Cancer cell membrane-coated nanoparticles for anticancer vaccination and drug delivery, *Nano Letters 14*, 2181–2188.

20. Alvarez-Erviti, L., et al., (2011). Delivery of siRNA to the mouse brain by systemic injection of targeted exosomes. *Nature Biotechnology, 29*, 341–345.

21. Hu, C. M. J., Fang, R. H., Copp, J., Luk, B. T., & Zhang, L., (2013). A biomimetic nanosponge that absorbs pore-forming toxins. *Nature Nanotechnology, 8*, 336–340.

22. Hu, C. M. J., et al., (2015). Nanoparticle biointerfacing by platelet membrane cloaking. *Nature, 526*, 18–121.

23. Hu, Q., et al., (2015). Anticancer platelet-mimicking nanovehicles. *Advanced Materials, 27*, 7043–7050.

24. Li, J., et al., (2016). Targeted drug delivery to circulating tumor cells via platelet membrane-functionalized particles. *Biomaterials, 76*, 52–65.

25. Dou, H., et al., (2006). Development of a macrophage-based nanoparticle platform for antiretroviral drug delivery. *Blood, 108*, 2827–2835.

26. Kim, H., Cohen, R. E., Hammond, P. T., & Irvine, D. J., (2006). Live lymphocyte arrays for biosensing. *Advanced Functional Materials, 16*, 1313–1323.

27. Huang, B., et al., (2015). Active targeting of chemotherapy to disseminated tumors using nanoparticle-carrying T-cells. *Science Translational Medicine, 7*, 291–294.

28. Morgan, R. A., et al., (2006). Cancer regression in patients after transfer of genetically engineered lymphocytes. *Science, 314*, 126–129.

29. Stephan, M. T., Moon, J. J., Um, S. H., Bershteyn, A., & Irvine, D. J., (2010). Therapeutic cell engineering with surface conjugated synthetic nanoparticles. *Nature Medicine, 16*, 1035–1041.

30. Dhal, P. K., et al., (2002). *Polymeric Drugs Encyclopedia Polymer Science Technology* (pp. 555–580). CWiley VCH: Weinheim, Germany.

31. Duncan, R., (2003). Polymer-drug conjugates. In: Budman, D., Calvert, H., & Rowinsky, E., (eds.), *Handbook of Anticancer Drug Development* (pp. 239–260). Lippincott, Williams, and Wilkins: Philadelphia, PA.

32. Harris, J. M., & Chess, R. B., (2003). Effect of pegylation on pharmaceuticals. *Nature Reviews Drug Discovery, 2*, 214–221.

33. Ringsdorf, H., (1975). Structure and properties of pharmacologically active polymers. *Journal of Polymer Science: Polymer Symposia, 51*, 135–153.

34. Davis, F. F., (2002). The origin of pegnology. *Advanced Drug Delivery Reviews, 54*, 457–458.

35. Kakizawa, Y., & Kataoka, K., (2002). Block copolymer micelles for delivery of gene and related compounds. *Advanced Drug Delivery Reviews, 54*, 203.

36. Pack D. W., et al., (2005). Design and development of polymers for gene delivery. *Nature Reviews Drug Discovery, 4*, 581–593.

37. Mankoff, D. A., (2007). A definition of molecular imaging. *The Journal of Nuclear Medicine, 48*, 18N–21N.

38. Hoffman, J. M., & Gambhir, S. S., (2007). Molecular imaging: The vision and opportunity for radiology in the future. *Radiology, 244*, 39–47. doi: 10.1148/radiol .2441060773.

39. Massoud, T. F., & Gambhir, S. S., (2007). Integrating noninvasive molecular imaging into molecular medicine: An evolving paradigm. *Trends in Molecular Medicine, 13,* 183–191. doi: 10.1016/j. molmed.2007.03.003.

40. Weissleder, R., & Pittet, M. J., (2008). Imaging in the era of molecular oncology. *Nature, 452,* 580–589. doi: 10.1038/nature06917.

41. Willmann, J. K., Bruggen, N. V., Dinkelborg, L. M., & Gambhir, S. S., (2008). Molecular imaging in drug development. *Nature Reviews Drug Discovery, 7,* 591–607. doi: 10.1038/nrd2290.

42. Herschman, H. R., (2003). Molecular imaging: Looking at problems, seeing solutions. *Science, 302,* 605–608. doi: 10.1126/ science.1090585.

43. Lammers, T., Kiessling, F., Hennink, W. E., & Storm, G., (2010). Nanotheranostics and image-guided drug delivery: Current concepts and future directions. *Molecular Pharmaceutics, 7,* 1899–1912. doi: 10.1021/ mp100228v.

44. Ntziachristos, V., Yoo, J. S., & Van Dam, G. M., (2010). Current concepts and future perspectives on surgical optical imaging in cancer. *The Journal of Biomedical Optics, 15,* 066024. doi: 10.1117/1.3523364.

45. Urano, Y., Sakabe, M., Kosaka, N., Ogawa, M., Mitsunaga, M., Asanuma, D., et al., (2011). Rapid cancer detection by topically spraying a γ-glutamyltranspeptidase-activated fluorescent probe. *Science Translational Medicine, 3,* 111–119. doi: 10.1126/ scitranslmed.3002823.

46. Frangioni, J. V., (2008). New technologies for human cancer imaging. *The Journal of Clinical Oncology, 26,* 4012–4021. doi: 10.1200/JCO.2007.14.3065.

47. Vande, W. C., & Oltenfreiter, R., (2006). Imaging probes targeting matrix metalloproteinases. *Cancer Biotherapy and Radiopharmaceutical, 21,* 409–417. doi: 10.1089/ cbr.2006.21.409.

48. Razgulin, A., Ma, N., & Rao, J., (2011). Strategies for *in vivo* imaging of enzyme activity: An overview and recent advances. *Chemical Society Reviews, 40,* 4186–4216. doi: 10.1039/C1CS15035A.

49. Finkel, T., Serrano, M., & Blasco, M. A., (2007). The common biology of cancer and ageing. *Nature, 448,* 767–774. doi: 10.1038/nature05985.

50. Zweier, J. L., & Hassan, T. M. A., (2006). The role of oxidants and free radicals in reperfusion injury. Cardiovascular Research, 70, 181–190. doi: 10.1016/j.cardiores. 2006.02.025.

51. Kobayashi, H., Ogawa, M., Alford, R., Choyke, P. L., & Urano, Y., (2010). New strategies for fluorescent probe design in medical diagnostic imaging. *Chemical Reviews, 110,* 2620–2640. doi: 10.1021/cr900263j.

52. Chang, M. C. Y., Pralle, A., Isacoff, E. Y., & Chang, C. J., (2004). A selective, cellpermeable optical probe for hydrogen peroxide in living cells. *The Journal of the American Chemical Society, 126,* 15392–15393. doi: 10.1021/ja0441716.

53. Miller, E. W., Albers, A. E., Pralle, A., Isacoff, E. Y., & Chang, C. J., (2005). Boronate-based fluorescent probes for imaging cellular hydrogen peroxide. *The Journal of the American Chemical Society, 127,* 16652–16659. doi: 10.1021/ja054474f.

54. Miller, E. W., Tulyathan, O., Isacoff, E. Y., & Chang, C. J., (2007). Molecular imaging of hydrogen peroxide produced for cell signaling. *Nature Chemical Biology, 3,* 263–267. doi: 10.1038/nchembio871.

55. Dickinson, B. C., & Chang, C. J., (2008). A targetable fluorescent probe for imaging hydrogen peroxide in the mitochondria of living cells. *The Journal of the American Chemical Society*, *130*, 11561. doi: 10.1021/ja802355u.

56. Abo, M., Urano, Y., Hanaoka, K., Terai, T., Komatsu, T., & Nagano, T., (2011). Development of a highly sensitive fluorescence probe for hydrogen peroxide. *The Journal of the American Chemical Society*, *133*, 10629–10637. doi: 10.1021/ ja203521e.

57. Dickinson, B. C., Huynh, C., & Chang, C. J., (2010). A palette of fluorescent probes with varying emission colors for imaging hydrogen peroxide signaling in living cells. *The Journal of the American Chemical Society*, *132*, 5906–5915. doi: 10.1021/ ja1014103.

58. Ye, Z., Chen, J., Wang, G., & Yuan, J., (2011). Development of a terbium complex based luminescent probe for imaging endogenous hydrogen peroxide generation in plant tissues. *Analytical Chemistry*, *83*, 4163–4169. doi: 10.1021/ac200438g.

59. Kim, G., Koo, L. Y. E., Xu, H., Philbert, M. A., & Kopelman, R., (2010). Nanoencapsulation method for high selectivity sensing of hydrogen peroxide inside live cells. *Analytical Chemistry*, *82*, 2165–2169. doi: 10.1021/ac9024544.

60. Lee, D., Khaja, S., Velasquez-Castano, J. C., Dasari, M., Sun, C., Petros, J., Taylor, W. R., & Murthy, N., (2007). *In vivo* imaging of hydrogen peroxide with chemiluminescent nanoparticles. *Nature Materials*, *6*, 765–769. doi: 10.1038/nmat1983.

61. Van de Bittner, G. C., Dubikovskaya, E. A., Bertozzi, C. R., & Chang, C. J., (2010). *In vivo* imaging of hydrogen peroxide production in a murine tumor model with a chemoselective bioluminescent reporter. *Proceedings of the National Academy of Sciences of the United States of America*, *107*, 21316–21321. doi: 10.1073/ pnas.1012864107.

62. Van de Bittner, G. C., Bertozzi, C. R., & Chang, C. J., (2013). A strategy for dualanalyte luciferin imaging: *In vivo* bioluminescence detection of hydrogen peroxide and caspase activity in a murine model of acute inflammation. *The Journal of American Chemical Society*, *135*, 1783–1795. doi: 10.1021/ ja309078t.

63. Weinstain, R., Savariar, E. N., Felsen, C. N., & Tsien, R. Y., (2014). *In vivo* targeting of hydrogen peroxide by activatable cell-penetrating peptides. *The Journal of American Chemical Society*, *136*, 874–877. doi: 10.1021/ja411547j.

64. Imamura, H., Huynh, N. K. P., Togawa, H., Saito, K., Iino, R., Kato-Yamada, Y., Nagai, T., & Noji, H., (2009). Visualization of ATP levels inside single living cells with fluorescence resonance energy transfer-based genetically encoded indicators. *Proceedings of the National Academy of Sciences of the United States of America*, *106*, 15651–15656. doi: 10.1073/pnas.0904764106.

65. Fehr, M., Lalonde, S., Lager, I., Wolff, M. W., & Frommer, W. B., (2003). *In vivo* imaging of the dynamics of glucose uptake in the cytosol of COS-7 cells by fluorescent nanosensors. *The Journal of Biological Chemistry*, *278*, 19127–19133. doi: 10.1074/jbc.M301333200.

66. Miyawaki, A., Llopis, J., Heim, R., McCaffery, J. M., Adams, J. A., Ikura, M., & Tsien, R. Y., (1997). Fluorescent indicators for $Ca^{2+}$ based on green fluorescent proteins and calmodulin. *Nature*, *388*, 882–887.

67. Vinkenborg, J. L., Nicolson, T., Bellomo, E. A., Koay, M. S., Rutter, G. A., & Merkx, M., (2009). Genetically encoded FRET sensors to monitor Zn2+ homeostasis in single cells. *Nature Methods*, *6*, 737–740. doi: 10.1038/ nmeth.1368.

68. Vinkenborg, J. L., Van Duijnhoven, S. M. J., & Merkx, M., (2011). Reengineering of a fluorescent zinc sensor protein yields the first genetically encoded cadmium probe. *Chemical Communications, 47,* 11879–11881. doi: 10. 1039/C1CC14944J.

69. Shi, H., He, X., Wang, K., Wu, X., Ye, X., Guo, Q., Tan, W., Qing, Z., Yang, X., & Zhou, B., (2011). Activatable aptamer probe for contrast-enhanced *in vivo* cancer imaging based on cell membrane protein-triggered conformation alternation. *Proceedings of the National Academy of Sciences of the United States of America, 108,* 3900–3905. doi: 10.1073/pnas.1016197108.

70. Yeh, H. Y., Yates, M. V., Mulchandani, A., & Chen, W., (2008). Visualizing the dynamics of viral replication in living cells via Tat peptide delivery of nuclease-resistant molecular beacons. *Proceedings of the National Academy of Sciences of the United States of America,* 17522–17525. doi: 10.1073/pnas.0807066105.

71. Shimizu, Y., Temma, T., Hara, I., Makino, A., Yamahara, R., Ozeki, E., Ono, M., & Saji, H., (2014). Micelle-based activatable probe for *in vivo* near-infrared optical imaging of cancer biomolecules. *Nanomedicine, 10,* 187–195. doi: 10.1016/j.nano.2013.06.009.

72. Tansi, F. L., Rüger, R., Rabenhold, M., Steiniger, F., Fahr, A., Kaiser, W. A., & Hilger, I., (2013). Liposomal encapsulation of a near-infrared fluorophore enhances fluorescence quenching and reliable whole body optical imaging upon activation *in vivo. Small, 9,* 3659–3669. doi: 10.1002/smll.201203211.

73. Tsuji, M., Ueda, S., Hirayama, T., Okuda, K., Sakaguchi, Y., Isono, A., & Nagasawa, H., (2013). FRET-based imaging of transbilayer movement of pepducin in living cells by novel intracellular bioreductively activatable fluorescent probes. *Organic Biomolecular Chemistry, 11,* 3030–3037. doi: 10.1039/c3ob27445d.

74. Alexander, V. M., Sano, K., Yu, Z., Nakajima, T., Choyke, P. L., Ptaszek, M., & Kobayashi, H., (2012). Galactosyl human serum albumin-NMP1 conjugate: A near infrared (NIR)-activatable fluorescence imaging agent to detect peritoneal ovarian cancer metastases. *Bioconjugate Chemistry, 23,* 1671–1679. doi: 10.1021/bc3002419.

75. Mitra, R. N., Doshi, M., Zhang, X., Tyus, J. C., Bengtsson, N., Fletcher, S., et al., (2012). An activatable multimodal/multifunctional nanoprobe for direct imaging of intracellular drug delivery. *Biomaterials, 33,* 1500–1508. doi: 10.1016/j.biomaterials. 2011.10.068.

76. Guiseppi-Elie, A., et al., (2002). A chemically synthesized artificial pancreas: Release of insulin from glucose-responsive hydrogels. *Advanced Materials, 14*(10), 741–746.

77. Wilson, A. N., et al., (2012). Bioactive hydrogels demonstrate mediated release of a chromophore by chymotrypsin. *Journal of Controlled Release.*

78. Fischel-Ghodsian, F., et al., (1988). Enzymatically controlled drug delivery. *Proceedings of the National Academy of Sciences, 85*(7), 2401–2406.

79. Yamaguchi, N., et al., (2007). Growth factor mediated assembly of cell receptor responsive hydrogels. *Journal of the American Chemical Society, 129*(11), 3041–3041.

80. Jin, R., et al., (2007). Enzyme-mediated fast *in situ* formation of hydrogels from dextrantyramine conjugates. *Biomaterials, 28*(18), 2791–2800.

81. Levesque, S. G., & Shoichet, M. S., (2007). Synthesis of enzyme-degradable, peptide-crosslinked dextran hydrogels. *Bioconjugate Chemistry, 18*(3), 871–885.

82. Miyata, T., et al., (1999). Preparation of an antigen-sensitive hydrogel using antigen-antibody bindings. *Macromolecules, 32*(6), 2081–2084.

83. Murakami, Y., & Maeda, M., (2005). DNA-Responsive hydrogels that can shrink or swell. *Biomacromolecules, 6*(6), 2921–2929.

84. Miyata, (2004). Preparation of reversibly glucose-responsive hydrogels by covalent immobilization of lectin in polymer networks having pendant glucose. *Journal of Biomaterials Science, Polymer Edition, 15,* 1081–1098.

85. Lutolf, M. P., et al., (2003). Synthetic matrix metalloproteinase-sensitive hydrogels for the conduction of tissue regeneration: Engineering cell-invasion characteristics. *Proceedings of the National Academy of Sciences, 100*(9), 5411–5418.

86. Gawel, K., et al., (2010). Responsive hydrogels for label-free signal transduction within biosensors. *Sensors, 10*(5), 4381–4409.

87. Lee, K. Y., & Mooney, D. J., (2001). Hydrogels for tissue engineering. *Chemical Reviews, 101*(7), 1861–1880.

88. Yu, H., & Grainger, D. W., (1993). Thermo-sensitive swelling behavior in crosslinked Nisopropylacrylamide networks: Cationic, anionic, and ampholytic hydrogels. *Journal of Applied Polymer Science, 49*(9), 1551–1563.

89. Suzuki, Y., et al., (1996). Change in phase transition behavior of an NIPA gel induced by solvent composition: Hydrophobic effect. *Polymer Gels and Networks, 4*(2), 121–142.

90. Ulijn, R. V., et al., (2007). Bioresponsive hydrogels. *Materials Today, 10*(4), 41–48.

91. Miyata, T., et al., (1999). A reversibly antigen-responsive hydrogel. *Nature, 399* (6738), 761–769.

92. Thornton, P. D., et al., (2007). Enzyme-responsive polymer hydrogel particles for controlled release. *Advanced Materials, 19*(9), 1251–1256.

93. Milton, J., et al., (1995). Dynamic feedback and the design of closed-loop drug delivery systems. *Journal of Biological Systems, 3,* 711–718.

94. Cohen, M. H., & Turnbull, D., (1959). Molecular transport in liquids and glasses. *The Journal of Chemical Physics, 31,* 1164.

95. Ravaine, V., et al., (2008). Chemically controlled closed-loop insulin delivery. *Journal of Controlled Release, 132*(1), 1–11.

96. Villoslada, P., et al., (2009). Systems biology and its application to the understanding of neurological diseases. *Annals of Neurology, 65*(2), 121–139.

97. Quo, C. F., et al., (2011). Adaptive control model reveals systematic feedback and key molecules in metabolic pathway regulation. *Journal of Computational Biology: A Journal of Computational Molecular Cell Biology, 18*(2), 161–182.

98. Hersel, U., et al., (2003). RGD modified polymers: Biomaterials for stimulated cell adhesion and beyond. *Biomaterials, 24*(240), 4381–4415.

99. Kim, S., & Healy K. E., (2003). Synthesis and characterization of injectable poly (Nisopropylacrylamide-co-acrylic acid) hydrogels with proteolytically degradable cross-links. *Biomacromolecules, 4*(5), 1211–1223.

100. Russell, R. J., et al., (1999). A Fluorescence-based glucose biosensor using concanavalin A and dextran encapsulated in a poly(ethylene glycol) hydrogel. *Analytical Chemistry, 71*(15), 3121–3132.

101. Deere, J., et al., (2002). Mechanistic and structural features of protein adsorption onto mesoporous silicates. *The Journal of Physical Chemistry B., 106*(29), 7341–7347.

102. Guiseppi-Elie, A., et al., (2002). Direct electron transfer of glucose oxidase on carbon nanotubes. *Nanotechnology, 13*(5), 559.

103. Shao, Y., et al., (2010). Graphene based electrochemical sensors and biosensors: A review. *Electroanalysis, 22*(10), 1021–1036.

# CHAPTER 6

# Bioresponsive Nanoparticles

DRASHTI PATHAK and DEEPA H. PATEL

*Department of Pharmaceutics, Parul Institute of Pharmacy and Research, Faculty of Pharmacy, Parul University, P.O. Limda, Ta. Waghodia, Vadodara–391760, Gujarat, India, Phone: 02668-260287, Fax: 02668-260201, E-mails: deepaben.patel@paruluniversity.ac.in, Pateldeepa18@yahoo.com (D. H. Patel)*

## 6.1 BIORESPONSIVE NANOPARTICLES

Homogeneous and heterogeneous nanoparticle (NP) assembly induced by ligand-specific immunorecognition is commonly used for biosensing applications. Nanomedicine is a research field that has recently made significant progress in the areas of drug delivery. A drug delivery vehicle must be able to carry a therapeutic cargo and be able to reach the target tissue and the intended intracellular target or compartment. This dissertation will focus on two major applications of drug delivery vehicles and novel improvements in these areas. The first area is vaccine delivery, where sensitive biological cargo must be delivered to activate immune cells.

A new peptide adjuvant called Hp91 was chosen to be co-delivered to dendritic cells (DCs) along with an antigen to ensure activation. The vaccine drug delivery vehicle designed to deliver these biomolecules consists of a poly(lactic-co-glycolic acid) (PLGA) core with the antigen encapsulated inside and the peptide adjuvant either co-encapsulated or conjugated to the outside. The PLGA successfully protected the protein and peptide from degradation, maintaining its activity. The vaccine delivery vehicles showed a marked improvement over free Hp91 in both human and mouse dendritic cell activation; with a maximum of 44 fold increase in IL-6 stimulation when the peptide was conjugated to the surface. The second area of research presented here is systemic drug delivery. For systemic delivery, long circulation is a desirable characteristic and poly(ethylene glycol) (PEG) is the current state of the art.

However, PEG presents two challenges:

1. PEG can hinder the nanoparticles from entering the cytosol across the endosomal or cellular membrane; and
2. PEG can sometimes induce immunogenicity resulting in accelerated blood clearance after repeated dosing.

A novel PEG shedding nanoparticle is presented as a solution to the first challenge. The ability to shed the PEG layer in response to a reduction in pH allows a fusogenic lipid layer to be exposed, which promotes membrane disruption. For patients who respond immunogenically to PEG, a biomimetic apolipoprotein coating to replace PEG altogether is presented. Apolipoprotein A1 is shown here to stabilize PLGA nanoparticles and increase their circulation half-life beyond that of the traditional PEG coating. These drug delivery vehicles were engineered from the foundation of a biodegradable, biocompatible PLGA polymer core with characteristic surface properties designed to overcome these specific challenges to drug delivery.

Nanotechnology has the potential to impact many long-standing challenges in medicine, such as selective drug delivery and sensitive detection of disease [1–3]. In recent years, mesoporous silica nanoparticles (MSNPs) have attracted attention as a promising component of multimodal nanoparticle systems. MSNPs are excellent candidates for many biomedical applications owing to their straightforward synthesis, tunable pore morphologies, facile functionalization chemistries, low-toxicity degradation pathways in the biological milieu, and capacity to carry disparate payloads (molecular drugs, proteins, other nanoparticles) within the porous core [1–7]

### 6.1.1   IMPROVING DRUG EFFICACY THROUGH DRUG DELIVERY

Improving drug efficacy is especially important in the fight against hard to treat diseases such as cancer, antibiotic-resistant bacterial infections, and chronic viral infections such as HIV and Hepatitis C. While new drug discovery efforts are invaluable, improving the efficacy of existing therapies can help turn the tide against these deadly diseases, resulting in millions of lives saved [7, 8]

For example, according to the National Breast Cancer Foundation, the fight against breast cancer is not a fight against a single disease, but a fight against no less than seven major subclasses of breast cancer (Ductal Carcinoma *In-situ*, Infiltrating Ductal Carcinoma, Medullary Carcinoma, Infiltrating Lobular Carcinoma, Tubular Carcinoma, Mucinous Carcinoma, and Inflammatory Breast Cancer). Each of these types of breast cancer are classified based on what type of tissue the cancer first develops in as well as physical characteristics of the tumors or cancer cells themselves.

In reality, each of these subclasses may be divided even further based on the actual genetic or epigenetic mutation that caused the cancer. The number of potential mutations in the approximately 420,000 genes in the human genome that might result in cancerous tumor formation is unknown at this time, but it is undoubtedly high. Each mutation represents a different disease, possibly requiring completely different therapeutic strategies, although the histology and presentation may be similar or even identical.

It becomes difficult to treat each genetic variation of cancer with a drug uniquely formulated specifically for that mutation.

Additionally, mutations may lead to drug resistance in the tumor cells, making the effective use of existing drugs even more challenging. By combining existing drugs with nanomedicine and drug delivery technologies, the number of possible tools available to the clinician treating cancer is multiplied and the chances of finding an effective formulation are greatly increased [9–11]. In addition, using proper drug delivery vehicles, multiple drugs can be delivered simultaneously, further increasing the available treatments for overcoming drug resistance in cancer tumors.

Similarly, in the fight against antibiotic-resistant microbes, the number of effective antibiotic drugs available is limited. Recently, infections from bacteria strains resistant to our last line of antibiotic defense, the carbapenem class of antibiotic drugs that includes vancomycin, have been increasing in number. There are currently no new drugs in the pipeline to combat carbapenem-resistant bacterial infections. The widespread emergence of such infections is considered a significant threat to the public health. Hospital-acquired infections of methicillin-resistant Staphylococcus aureus (MRSA) were responsible for the deaths of 19,000 Americans in 2005 (the last year for which the US Centers for Disease Control has recorded). In order to prolong the lifespan of these critical, life-saving drugs, more efficient and effective delivery using nanomedicine technologies must be explored in earnest. Vaccination is another strategy that can benefit from the use of drug delivery vehicles [12–14]. Vaccinations rely on the body's own defenses to combat the infection by providing the immune system with the proper tools to identify the infection. In one immunological pathway, antigens from the infectious pathogen are taken up by DCs, professional antigen-presenting cells, which then present the antigens to other immune cells such as T cells and B cells.

In this function, DCs act as messengers between the innate and the adaptive immune system, delivering the necessary information for pathogen identification [15, 16]. Drug delivery vehicles that are taken up more efficiently by the professional antigen-presenting immune cells can improve the efficacy of the vaccine [17–19].

Vaccines have traditionally incorporated adjuvants that increase the immunologic activity but have no antigenic activity themselves. The use of adjuvants allows the immune system to respond more strongly, increasing immunity in the recipient. Adjuvants do this by acting as synthetic versions of evolutionarily conserved molecules called pathogen-associated molecular patterns, or PAMPs [15].

Without the use of adjuvants, the innate immune system is not activated and immunity is conferred at a much lower level, if at all. This complex series of interactions is a fertile ground for improvements through the use of drug delivery vehicles. Immune cells preferentially engulf particulates of a certain size [20, 21].

Using a drug delivery vehicle allows the selection of particulate size independent of the antigen or adjuvant molecules being delivered [22]. Additionally, co-packaging antigens and adjuvants into the same drug delivery vehicle can ensure that any immune cell that takes up the vehicle will be fully activated [23–25].

This can improve the effectiveness of a vaccine and increase conferred immunity [26]. New strategies are also being explored that combine personalized medicine and vaccination strategies to induce an immune response against cancerous tumors [27, 28]

Drug delivery vehicles will play a key role in this emerging field of personalized medicine by allowing tumor-associated molecules unique to the patient to be reliably and reproducibly delivered to the immune system. Careful attention needs to be paid to avoid overstimulation of the immune system, which can lead to autoimmunity.

## 6.2 PLGA NANOPARTICLES AS VACCINE DRUG DELIVERY VEHICLE

Vaccination remains the most successful prophylaxis for infectious disease and in the past decades, it has also been explored as an approach to prevent or cure cancer [27, 28]. Since peptides tend to be unstable *in vivo*, NPs can protect them from degradation and potentially increase the immune response to peptide and possibly protein vaccines. Encapsulation of antigen peptides into biodegradable spheres has been shown to increase MHC-class-I presentation. Delivery of OVA protein as antigen in poly-g-Glutamic acid nanoparticles (gPGA-NPs) lead to increased immune responses in comparison to vaccinations using the same amount of free OVA protein. DCs are the most potent antigen-presenting cells and central for the initiation of adaptive immune responses. DCs need to receive a maturation signal in order to present antigen, upregulate co-stimulatory and adhesion molecules, and become potent activators of T cells. Antigen-displaying mature DCs can then activate T cells to act as CTLs. Collaborating researchers previously identified several immunostimulatory peptides (ISPs), derived from the endogenous protein HMGB1, which can activate both mouse and human DCs [29].

In addition, polymeric NPs are more stable in the gastrointestinal tract as compared to other carriers like liposomes and can be used for oral vaccine development [30].

Different composition polymers allow for controlled and prolonged release of cargo, allowing for antigen depot formation at the injection site, again another major advantage for vaccine development.

## 6.3 DRUG DELIVERY VEHICLE WITH PH TRIGGERED PEG SHEDDING

Nanoparticle drug delivery vehicles with peptides conjugated on the outside present a bioactive surface that is efficient at interacting with immune cells. The surface of the drug delivery vehicle is important in interactions with cells other than immune cells. Most drug delivery vehicles have the ultimate goal of delivering their cargo to the interior of a cell. A nanoengineered surface coating becomes invaluable if it can facilitate the delivery of the cargo into the desired intracellular compartment [31].

The ability to deliver drugs more effectively and efficiently to the site of interest translates into less harmful systemic side effects and more beneficial therapeutic action. This ability is particularly useful in the fight against cancer, where harmful side effects limit the tolerable dose of chemotherapeutics. Using polyethylene glycol (PEG) has become a popular strategy to create long-circulating drug delivery nanoparticles and other vehicles by reducing protein adsorption, macrophage uptake, and particle aggregation, thus increasing systemic circulation lifetime. Although useful in increasing circulation half-life, the PEG layer may become a detriment upon reaching the target tissue, hindering the entry of the nanoparticles into the cell, or preventing its escape from the endosome after being endocytosed [30].

A drug delivery vehicle with pH-sensitive PEG shedding would be especially useful in cancer drug delivery by exploiting the slightly acidic extracellular space of tumors (around pH ~ 6.5). Upon arrival at the tumor site, a correctly tuned pH-sensitive particle would be able to shed its PEG coating, thus enabling it to fuse with the cell membrane and be internalized. Additionally, nanoparticles taken up via the endosomal pathway can be tuned to lose the protective PEG coating upon acidification of the late endosome or early lysosome. Fusing with the endosomal membrane then becomes possible and escape from the endosome can be achieved. The importance of endosomal escape cannot be underestimated, especially for the delivery of degradable payloads like siRNA and other biologics that are typically degraded inside the highly acidic lysosome. Previous research has shown PEG shedding to improve intracellular drug delivery using polymersomes and polyplex micelles. Other PEG shedding molecules that rely on the reduction of disulfide bonds have been used in liposomes and lipoplexes to some success. More recently, Gao et al. Have demonstrated a technique to directly observe PEG shedding using a pair of dye and quencher, confirming the benefits of PEG shedding to intracellular delivery [28, 29].

PEG is the current state of the art in long-circulating nanoparticle technologies. In the previous chapter, we discussed an improvement upon the PEG coating by making the PEG sheddable in response to an environmental trigger. This is particularly useful for the delivery of sensitive biomolecules. However, there is a need for an alternative to PEG in order to increase the number of tools available to drug delivery vehicle designers. For systemic delivery of therapeutic or diagnostic agents, clearance from the bloodstream is one of the most challenging hurdles to overcome [30]. Drug delivery vehicles need to lie within a sweet spot of physical characteristics including size, surface charge or zeta-potential, solubility, and steric stability in order to navigate the bodies' numerous defenses against foreign material and pathogens. Apart from size, most, if not all, of the optimal characteristics are governed by the nanoparticles' surface properties. Charged surface molecules dominate the zeta potential. Solubility can be altered through the use of a surfactant or amphiphilic molecules that alter the polarity or hydrophobicity at the nanoparticle surface. Steric stability is likewise a property conferred by surface molecules which prevent aggregation and adsorption of proteins. PEG performs well in conferring desirable surface properties to nanoparticles in all these categories. As useful as PEG has been proven to be in creating long-circulating nanoparticles, alternatives should be explored. Just as the overuse of a single antibiotic can lead to resistant pathogens, relying on a single technology to solve a problem within a complex and adaptive system such as the human body can lead to problems down the road. Recent studies have shown that over 25% of the population already produces anti-PEG antibodies which are no surprise given the ubiquitous nature of PEG in everyday products such as toothpaste and shampoo. The presence of anti-PEG antibodies is also strongly correlated with the rapid clearance of certain PEGylated biomolecules. Lipid-polymer hybrid nanoparticles have been shown to be a suitable long-circulating drug delivery vehicle platform [27]. But the typical lipid-polymer hybrid nanoparticles formulation relies on PEG to impart long circulation properties. In the search for an alternative to PEG that could be used with the lipid-polymer hybrid platform, we turned to nature for a solution. The body naturally produces hybrid nanoparticles called high-density lipoproteins (HDL) and low-density lipoproteins (LDL) that circulate in the bloodstream and transport fats and cholesterol through the body. These natural particles are composed of a hydrophobic core made up of cholesterol, triglycerides, fats, and fatty acids.

## 6.4 BIOLOGICAL BARRIERS AND STRATEGIES FOR BIORESPONSIVE NANOPARTICLES

The complexity and heterogeneity of the human body and, particularly, the tumor pathophysiology, together with an incomplete comprehension of the nano-bio interactions, represent main hurdles for the establishment of nanomedicines as a new paradigm in cancer therapy. Despite considerable efforts on developing nanomedicines for non-invasive administration (e.g., oral, pulmonary, nasal, and transdermal delivery routes), most cancer nanomedicines are envisioned to be administered systemically. After intravenous injection, the nanoparticles-based therapeutics faces the challenge to reach and accumulate at the targeted tumor site, in order to exert a therapeutic effect. Therefore, the *in vivo* PK and related therapeutic efficacy of newly developed nanomedicines crucially depends on their capability to overcome multiple biological barriers. Generally, the passive tumor accumulation and localization of nanomedicines are favored by an increased leakiness of the abnormal tumor microvasculature and defective lymphatic drainage, enabling the nanoparticulate systems to extravasate from the blood circulation into the perivascular tumor microenvironment (TME) and to be retained within the tumor tissue [31–34].

Once a nanoparticle enters a biological system, such as the blood, interstitial fluid, or extracellular matrix (ECM), it is exposed and interacts with a variety of biomolecules, particularly proteins, which are tissue or organ-specific in terms of their chemical and biological composition. The phenomena of protein adsorption and protein corona formation, as well as their composition, are highly dependent on the physicochemical properties of the nanomaterials (i.e., size, morphology, chemical composition, and surface chemistry). The composition of 11 biological system (blood, interstitial fluid, TME, intracellular compartment, etc.), the pathological state, and other factors such as temperature, pH, dynamic sheer stress, and exposure time. In turn, the protein corona formation alters the nanoparticles size, surface properties, stability, and functionality, thus providing the nanomedicines with a new entity that significantly impacts their biocompatibility, PK profile, biodistribution, tumor cellular internalization, intracellular trafficking, drug release and, consequently, their safety and therapeutic efficacy. For example, the surface adsorption of opsonins can induce the rapid recognition and phagocytosis by the mononuclear phagocytic system (MPS), resulting in the clearance of the nanoparticles from the systemic circulation and the accumulation in the MPS associated organs (i.e., liver, and spleen). Contrarily, the surface binding of dyopsonins, such as apolipoproteins and albumin, can render stealthy properties to the nanoparticles, avoiding the opsonization by the MPS In addition; the formation of a protein layer on the nanoparticles surface can affect their non-specific cellular interactions, uptake, and the intracellular trafficking. As an example, the cellular internalization of silica nanoparticles revealed to be considerably more efficient when the particles were incubated with the cells in serum-free conditions, since the proteins bound on the nanoparticles surface significantly decreased their adhesion to the cellular membrane and, consequently, lowered their cellular uptake. However, the formation of a protein layer also diminishes the interaction with the cellular milieu, attenuating the acute cytotoxic effect of nanomaterials. In the case of nanoparticles functionalized with targeting ligands, the formation of a protein corona can affect the function of conjugated targeting moieties by displacing, altering their orientation, disrupting their structure and conformation, or masking their recognition, thus limiting the specific interaction and internalization of the nanomedicines by the targeted cancer cells and, consequently, impacting their targeting efficiency and biofate. Contrarily, the protein corona composition can also be designed for improving the targeting of nanotherapeutics. For instance, the adsorption of apolipoprotein E has recently been shown to drive the *in vivo* targeting of siRNA lipoplexes to hepatocytes. Similarly, the surface modification of gold (Au) nanoparticles with apolipoprotein E and albumin has been demonstrated to prolong their blood circulation time, and to significantly increase their translocation into the brain and accumulation in the lungs. Furthermore, the protein corona can modify the dissolution rate of nanoparticles and, consequently, the release profiles of the therapeutic cargo.

## 6.5 OPSONIZATION AND DEGRADATION OF BIORESPONSIVE NANOPARTICLES

Nanomedicines delivered systemically by intravenous injection are immediately subjected to rapid clearance from the blood circulation. Although the eventual clearance and biodegradation of nanomedicines are important prerequisites from a toxicological perspective, their rapid removal from the bloodstream can result in a primary accumulation in organs like the liver, spleen, and kidneys. This unfavorable PK and biodistribution between the targeted tumor site and other tissues can not only offset the desired therapeutic effect and dramatically impair the therapeutic efficacy of nanomedicines, but also induce off-target toxicity. Therefore, the rational design of nanomedicines that enable to circumvent the mechanisms of rapid clearance, to achieve a prolonged circulatory half-life and to preferably accumulate at the tumor site, are determinant for the overall effectiveness of nanoparticle-based therapeutics, and represent some of the main focuses in cancer nanomedicine research. The MPS and direct renal filtration are the two major physiological mechanisms responsible for the clearance of nanoparticles from the systemic circulation. The MPS, also known as reticuloendothelial system (RES), is composed of phagocytic cells, including macrophages, monocytes, and Kupffer cells that reside in MPS associated organs, such as the liver, spleen, lymph nodes, and bone marrow, which are responsible for engulfing and eliminating external organisms, viruses, and particles traveling in the blood circulation. When entering the bloodstream, nanoparticles are immediately opsonized by plasma proteins, typically albumin, immunoglobulins, complement proteins, and apolipoproteins, resulting in the formation of a protein corona onto the nanoparticles' surface. The opsonization promotes the recognition and phagocytic clearance of the nanoparticles by the MPS, through the binding of the adsorbed opsonins to specific receptors expressed on the phagocytic cells, followed by enzymatic degradation. The nanoparticles that are not degradable by enzymatic breakdowns, such as inorganic nanoparticles, will be transported by the phagocytes and accumulate in the liver and spleen. In addition to the association with the MPS, the intrinsic physiology of the liver and spleen can contribute for the permanent excretion of nanomedicines from the body. In the liver, hepatocytes may endocytose and slowly degrade the nanoparticles, subsequently eliminating them in the biliary system. In turn, nanoparticulate systems with sizes larger than 200 nm and long blood half-lives can be physically filtered from the bloodstream through the blood filtration system of the spleen. The clearance of nanomedicines by the MPS is complemented by the renal filtration system. The glomerular bed of the kidneys is characterized by a fenestrated capillary epithelium with an adjacent basement membrane and an epithelial layer of podocytes, which present a combined physiological pore size of approximately 4.5–5 nm. Contrarily to the active phagocytic role of the MPS, the renal clearance is fundamentally a passive mechanism, which is predominantly affected by the particle size rather than the surface properties of nanomedicines. Moreover, upon renal filtration, nanoparticles are directly excreted from the body in the urine, instead of accumulating in the kidneys. The clearance process and *in vivo* biofate of nanomedicines are generally influenced by their physicochemical properties, particularly the size, shape, surface charge, and hydrophobicity/hydrophilicity. In general, spherical nanoparticles with a hydrodynamic diameter smaller than 6 nm are rapidly dialyzed from the blood circulation through the renal filtration system, independently of their surface charge, while the renal clearance of 6–8 nm-sized nanoparticles significantly depends on their surface properties. Accordingly, positively charged nanoparticles smaller than 8 nm have exhibited greater glomerular filtration than the negatively charged and neutral counterparts with similar dimensions, due to an increased interaction with the anionic moieties of the glomerular capillary wall. On the contrary, 13 nanomedicines larger than 8 nm cannot be handled by the kidneys and tend to be cleared by the MPS. As mentioned above, the phenomena of opsonization and protein corona formation play critical roles in the process of MPS clearance. In this regard, ionic, and/or hydrophobic nanoparticles present a higher propensity to be opsonized and phagocytized by the MPS cells, compared to their neutral and hydrophilic counterparts, which minimally interact with the plasma proteins, owing to their sterically stabilized surface. Therefore, extensive efforts have been devoted to rationally engineer the surfaces of nanomedicines, with the aim of mitigating their opsonization and MPS clearance, prolonging their blood half-life, and enhancing their tumor accumulation. The most widely explored strategy to achieve these goals consists of adsorbing or conjugating polyethylene glycol (PEG) or PEG derivatives onto the nanoparticles' surface, a process also known as pegylation. The stealth properties of PEG arise from its hydrophilicity, neutral charge, and steric repulsion, forming a hydrating shell on the nanoparticles' surface that minimizes the interaction with plasma proteins and, consequently, the recognition and internalization by the MPS cells. As a result, pegylated nanoparticles normally travel longer in the systemic circulation, giving an opportunity to accumulate more efficiently in the tumor tissues. Additionally to PEG-based moieties, other hydrophilic polymers, including polyacrylamide, poly(vinyl alcohol), poly(N-vinyl-2-pyrrolidone) and polysaccharides, such as dextran, heparin, chitosan (CS), and hyaluronic acid (HA), have demonstrated similar stealth features and provided significant benefits on prolonging the blood half-lives of nanomedicines.

## 6.6  POROUS SILICON PARTICLES

PSi was first reported by Arthur Uhlir Jr. and Ingeborg Uhlir in 1956, at the Bell Laboratories, but it was only 15 years after that the porous crystalline silicon (Si) structure of this material was described. Thereafter, considerably growing attention has been drawn to PSi, particularly due to the remarkable work of Prof. Leigh Canham, who discovered the photolumi-nescence of highly porous Si wires, owing to the two-dimensional quantum size effect and, later on, demonstrated the *in vitro* biocompatibility and bioactive properties of PSi. These findings are considered the milestones for the following application of PSi materials in biomedicine, particularly in the field of drug delivery. Over the last decade, PSi nanoparticles have been extensively explored as a nano tool for biomedical applications, due to significant advances in their fabrication and surface modification methods, the manipulation and fine-tuning of their physicochemical properties, as well as the increasing comprehension of their interactions with biological systems. In this regard, PSi nanoparticles present advantageous physico-chemical and biological features, including a high surface-to-volume ratio, large surface area (300–1000 $m^2.g^{-1}$) and pore volume (0.9 $cm^3.g^{-1}$), high chemical, mechanical, and thermal stability, as well as superior biocom-patibility and biodegradability. Additionally, the top-down fabrication method of these nanoparticles enables their scaled-up production and an easy control over their particle and pore sizes, depending on the fabrica-tion parameters applied. Moreover, the surface of PSi nanoparticles can be straightforwardly modified with different functional groups, and further functionalized with numerous polymers and biomolecules, in order to attain control over the release of therapeutic cargos, target specific organs, tissues, and cells, or improve the biological performance.

### 6.6.1 PSi PARTICLES FOR DRUG DELIVERY

In addition to the superior biocompatibility and suitable biodegradability of PSi particles, the unique nanoporous structure, large pore volume and the high surface area-to-volume ratio of these platforms render them an ideal candidate for drug delivery purposes. Different types of therapeutic cargos, including small drug molecules, proteins, peptides, and nucleic acids can be efficiently loaded into the porous network of PSi carriers, and released in a controlled mode by pore diffusion or upon the dissolu-tion of the PSi matrix. The drug loading and release profiles of molecules incorporated into the pores of PSi are governed by the physicochemical properties of the PSi material, such as the pore size and volume, surface charge, chemistry, and modification, hydrophobicity/hydrophilicity and its degradation rate. In addition, the physicochemical properties of the payload, as well as the solvent, method, and technical parameters used for drug loading are determinant for the efficiency and reproducibility of the loading process and for attaining the intended release behavior. Different methods can be applied for loading active molecules into the pores of PSi materials, including immersion, covalent grafting, impregnation, and drug entrapment by oxidation. Among these techniques, the immersion method is the most frequently used, owing to its straightforwardness, the feasibility to load a wide range of molecules with distinct physicochemical properties, and the possibility to be carried out in mild chemical conditions and at room temperature, thus being suitable for loading labile compounds, such as biomolecules. This method consists of the simple immersion of the PSi particles in a drug solution and relies on the diffusion of this solution into the porous structure of PSi and subsequent physical adsorption of the drug molecules to the pore surface. The phenomenon of physical adsorption is driven by the spatial confinement and chemical interactions established between the loaded molecules and the PSi pore surface. Therefore, it is highly dependent on the physicochemical characteristics of the PSi particles, namely the pore size, surface chemistry, and hydrophilicity/hydrophobicity, the intrinsic properties of the loaded molecules, the surface tension, viscosity, and concentration of the drug loading solution, the loading time, and the temperature. Consequently, all these parameters have a significant impact on both the drug loading efficiency and reproducibility, and the release kinetics of the loaded molecules from the PSi matrix. An alternative and more robust approach for drug loading into PSi carriers involve the covalent grafting of the drug molecules to the inner and/or outer surface of the PSi materials. In this case, the release of the active payload from the PSi particles only occurs after cleavage of the covalent bonds of the PSi-drug conjugates or degradation of the PSi matrix. Therefore, this methodology enables a precise control over the drug loading and release kinetics by tuning the type and conditions of the coupling chemistry applied and/or the degradation rate of the PSi framework. Although this approach grants more reproducible loading and releases profiles in comparison to the physical adsorption method, the attainable drug loading degree is generally inferior, since it depends on the number of functional groups available for drug conjugation at the PSi surface. This method was applied, for example, for conjugating methotrexate (MTX) to the pore surfaces of PSi nanoparticles, resulting in a prolonged release of this anticancer agent for up to 96 h and enhanced *in vitro* antiproliferative effect. In addition, an increased retention of the drug molecules inside the PSi materials and, consequently, a sustained drug release can be achieved by oxidizing the surface of the PSi pores after drug loading. During the oxidation process, additional oxygen atoms are inserted in the PSi matrix, leading to a reduction of the pore volume and an efficient entrapment of the active payload inside the PSi pores. Following this approach, two compounds, cobinamide, and rhodamine B, were co-loaded with an oxidizing agent, sodium nitrite, inside freshly etched PSi films, which were subsequently fractured into PSi micropar-ticles. The oxidation-mediated trapping of the model compounds into the PSi material showed, not only 33 to improve the drug loading degree, but also and most importantly, to remarkably sustain the drug release by 20-fold, compared to the pre-loading oxidized counterparts. One of the main challenges and focuses in the field of drug delivery, particularly for cancer therapy, is associated with the poor aqueous solubility of drugs and consequent limited bioavailability. The formulation of poorly water-soluble drugs into PSi particulate systems has been demonstrated to significantly increase their dissolution in aqueous media. When loaded in the nanosized pores of PSi particles, the drug molecules are spatially confined and restricted from rearranging into three-dimensional crystal lattices, thus remaining stable in an amorphous state or forming nanocrys-tals, which ultimately results in a higher dissolution rate, compared to the bulk counterparts. Despite the above mentioned advantageous features of PSi particles for loading and delivering the therapeutic agents, the degra-dation and off-target uncontrolled release of the loaded cargo, owing to the unrestrained access of the release media and biological fluids to the open pores of the PSi matrix, are major drawbacks limiting the applica-tion of these biomaterials for drug delivery. Therefore, different strate-gies have been explored for circumventing this problem and attaining control over the drug release kinetics from the PSi particles, including the loading of the drug molecules by covalent grafting or oxidation induced trapping previously described in this section. The physical capping of the pore apertures, the chemical grafting of pore gating and stimuli-responsive systems, or the encapsulation of the PSi particles within other carriers. The drug diffusion process from PSi-based drug delivery systems can be modified by physically adsorpting or covalently conjugating biocompat-ible and biodegradable nonresponsive polymers or lipids to the surface of PSi particles. This strategy has been applied, for example, by coating the surface of PSi particles with CS for prolonging the release of insulin and enhancing insulin permeation across an *in vitro* intestinal monolayer model. In another study, a solid-lipid nanocomposite combining glycerol monostearate and phosphatidylcholine was deposited onto the surface of PSi nanoparticles, improving their cytocompatibility, colloidal dispersity, and *in vitro* stability in human plasma, and sustaining the release of the encapsulated furosemide. Furthermore, the surface of PSi particles can be functionalized with different stimuli-responsive gate-keeping systems, in order to attain a spatiotemporal control over the drug release kinetics. In this regard, a smart PSi-based nanocomposite, envisioned for sequential combination cancer therapy, was engineered by combining PSi nanoparticles and a pH-responsive nanovalve system. This dual-drug delivery nanoplatform was designed by covalently conjugating a pH-responsive cationic polymer, poly(β-amino ester), to the surface of DOX-loaded PSi nanoparticles, followed by stabilization with PTX-encapsulating micelles composed by a PEG and poly(propylene glycol) triblock copolymer. As a result, a remarkable synergistic chemotherapeutic effect was achieved *in vitro*, through an immediate release of PTX from the polymeric micelles, followed by a pH-triggered release of DOX from the PSi nanoparticles.

### 6.6.2  *PSi PARTICLES FOR ACTIVE CANCER TARGETING*

Although PSi-based particulate systems have been demonstrating a high promise for drug delivery purposes, the biosafety and therapeutic efficiency of PSi-based nanomedicines, particularly when considering cancer therapy, is highly dependent on their capacity to accumulate in the malignant tissue, specifically attach to the targeted cells, and deliver the therapeutic cargo in a sufficient concentration at the site of action. Therefore, one of the major focuses with respect to the application of PSi nanoparticles in cancer nanomedicine has been to develop active targeting strategies to selectively guide these nanovehicles to the surface of the cancer cells and trigger the receptor-mediated endocytosis locally. In this context, the surface chemical versatility of the PSi nanoparticles represents one of the most attractive features, for enabling the surface functionalization with a variety of targeting ligands, such as tumor homing peptides, targeting antibodies, and DNA aptamers, 261 thus tremendously expanding the potential of this nanoplatforms for cancer drug delivery and therapeutic purposes. In this regard, the surface of THCPSi nanoparticles was decorated with a tumor homing peptide targeting the mammary-derived growth inhibitor

(MDGI) receptor. After intravenously administered into nude mice bearing subcutaneous MDGI-expressing tumors, the targeted nanoparticles showed an approximate 9-fold higher accumulation at the tumor site, in comparison to the non-functionalized particles. In another approach, a straightforward and efficient method based on copper-free click chemistry was applied to covalently conjugate RGD and iRGD tumor penetrating peptides on the surface of APSTCPSi nanoparticles for targeting the tumor neovasculature. Both RGD and iRGD modified PSi nanocarriers exhibited a significantly enhanced cellular uptake in EA.hy926 endothelial cells *in vitro*, when compared to the unmodified APSTCPSi nanoparticles.205 In addition to tumor homing peptides, the surface of PSi nanocarriers can be functionalized with targeting antibodies that are specific to different tumors. For example, MLR2, mAb528, and Rituximab antibodies have been successfully conjugated onto the surface of PSi nanoparticles to specifically target neuroblastoma, glioblastoma, and B lymphoma cells, respectively. Furthermore, size- and shape-controlled PSi nanodiscs were recently fabricated by a new method combining colloidal lithography and metal-assisted chemical etching, loaded with an anticancer agent, camptothecin, and further modified with a MLR2 anti-p75 antibody for targeting the p75NTR neurotrophin receptor expressed on the surface of neuroblastoma cells (SH-SY5Y). After antibody functionalization, the drug-loaded nanodiscs were found to be selectively attached and killed the cancer cells. Alternatively, the tumor homing of PSi nanovectors can be attained by exploring biological targets overexpressed in stroma cells. Accordingly, a Ly6C antibody was conjugated on the surface of PSi nanoparticles as a dual-targeting to pancreatic tumor-associated endothelial cells and macrophages. Contrarily to the control nanocarriers, the Ly6C antibody decorated PSi nanoparticles exhibited high affinity to the cells expressing Ly6C *in vitro*. Moreover, the targeted nanocarriers were shown to accumulate in the tumor-associated endothelial cells within 15 minutes, after their intravenous injection in orthotopic human pancreatic cancer-bearing nude mice. Interestingly, after extravasation through the endothelial cell monolayer, the targeted PSi nanoparticles were engulfed by the Ly6C expressing tumor-associated macrophages.

## 6.7 NANOPARTICLES ACTIVATION BY HYPOXIA

Due to the central role of hypoxia in enhancing tumor angiogenesis, metastasis, epithelial to mesenchymal transition, tumor invasiveness, and suppression of immune reactivity occurs. There has arisen great interest in the development of nanoparticles that can target the hypoxic regions within the tumor. For example, He et al. reported the fabrication of dual-sensitive nanoparticles with hypoxia and photo-triggered release of the anticancer drug. The authors developed dual stimuli nanoparticles through the self-assembly of polyethyleneimine-nitroimidazole micelles (PEI-NI) further co-assembled with Ce6-linked hyaluronic acid (HC). Hypoxia-mediated activation was achieved by the incorporation of nitroimidazole (NI), a hypoxia-responsive electron acceptor. Hydrophobic NI segments would be converted to hydrophilic 2-aminoimidazole under hypoxic conditions, thereby aiding in the release of the anticancer drug (doxorubicin (DOX)) loaded inside the nanoparticles. Another hypoxia-sensitive moiety is the azobenzene (AZO) group. The AZO group was introduced between the polyethylene glycol (PEG) and PEI for the construction of nanocarrier for the delivery of siRNA [44]. When these particles entered into hypoxic TME, the AZO bond was cleaved to trigger de-shielding of the PEG coating and the subsequent release of PEI/ siRNA nanoparticles. The exposed positive charge on the particles further facilitated the enhanced cellular uptake of PEI/siRNA nanoparticles. Xie et al. reported the development of hypoxia-responsive nanoparticles for the codelivery of siRNA and DOX. In this study, polyamidoamine (PAMAM) dendrimer was conjugated to PEG using AZO, which is a hypoxia-sensitive linker to form PAMAM-AZO-PEG (PAP). DOX was loaded into the hydrophobic core of PAMAM, and hypoxia-inducible factor 1a (HIF-1a) siRNA was electrostatically loaded onto the surface of PAMAM through ionic interactions between the anionic siRNA and amine groups of PAMAM. The PEG in PAP would prevent the nanoparticles from opsonization and prolong their circulation time in the blood. Upon reaching the tumor and exposure to hypoxic TME, PEG groups would be detached from the PAMAM surface due to the breakage of the AZO group to amino aromatics, causing the exposure of positively charged PAMAM. Once PAMAM has been taken up by tumor cells, PAMAM escapes from endosomes through the proton pump effect and releases the DOX and HIF-1a siRNA. Yang et al. have reported the one-pot synthesis of hollow silica nanoparticles encapsulated with catalase (CAT) and Ce6 doped into the silica lattice. CAT is a water-soluble $H_2O_2$ decomposing enzyme which triggers the decomposition of $H_2O_2$ to $H_2O$ and $O_2$. The nanoparticles were further modified with mitochondrial targeting moiety ((3-carboxypropyl) triphenylphosphonium bromide (CTPP)) and pH-responsive charge convertible polymer through electrostatic interaction. Upon reaching acidic TME, the polymeric coating would undergo charge conversion from negative to positive, thereby enhancing the cellular internalization. The mitochondrial targeting moiety helps in enhancing photodynamic therapy-induced cell death and the CAT encapsulated inside would decompose the tumor endogenous $H_2O_2$, thereby overcoming hypoxic environment in the tumor and enhancing the photodynamic therapy of solid tumors. These types of smart nanoparticles can overcome the limitations of conventional photodynamic therapy. Despite the advances in the development of hypoxia-responsive nanoparticles, getting these nanoparticles into hypoxic region is quite challenging as these regions are typically located deep inside the tumor with less vasculature, where the mass transport is through diffusion. For most of the nanoparticle systems, the diffusion rate would be insufficient within solid tumors and hence nanocarriers with a higher diffusion rate of small molecules would be a better option for carrying and releasing hypoxia-activated prodrugs within TME.

## 6.8   NANOPARTICLES ACTIVATION BY ENZYMES

TME also have upregulated levels of enzymes such as matrix metalloproteinase (MMP), which is predominantly a Fabrication of DOX-loaded PEI-NI-based nanoparticle co-assembled with HA-Ce6, (b) CD 44-mediated endocytosis and release of DOX in response to hypoxia generated by laser irradiation involved in tumor development and proliferation. The nanoprobe was constructed through the self-assembly of hexahistidine-tagged (His-Tagged) fluorescent protein and nickel ferrite nanoparticles. The nickel ferrite nanoparticles functioned as protein binders of His-Tagged fluorescent protein and fluorescent quencher. The nanoprobe was reported to be turned on by the presence of MMP-2, leading to enhanced cellular uptake and the restoration of fluorescence, thereby enabling the visualization of nanoparticles within tumor tissue. Ma et al. reported the fabrication of polymeric conjugate for mitochondrial targeting for paclitaxel (PTX) delivery. The polymeric conjugate consists of a PAMAM-based dendrimer core into which triphenylphosphine and PTX were conjugated through an amino bond and disulfide bonds, respectively. To enhance the circulation time of the polymeric conjugate in the blood, PEG was conjugated via the MMP-2 sensitive peptide

(GPLGIAGQ). The conjugates accumulate in tumor tissue through the EPR effect. Once the conjugate enters tumor cells, the PEG layer is detached from PAMAM by cleavage of MMP-2 sensitive peptide by the action of MMP-2. The conjugate would then target the mitochondria via triphenylphosphine and PTX would be released in the cytoplasm. Ansari et al. have reported the synthesis of theranostic nanoparticles that possess enzyme-specific drug release and *in vivo* magnetic resonance imaging (MRI). The nanoparticles were synthesized through the conjugation of ferumoxytol (FDA approved iron oxide nanoparticles) to MMP-14 activatable peptide conjugated to azademetylcolchicine (ICT) (CLIO-ICT). Upon reaching the tumor, the CLIO-ICT would be converted from non-toxic form to toxic form by the action of MMP-14 thereby, releasing potent ICT. This type of nanoparticles also enables the real-time monitoring of accumulation and localization of drug at the tumor site through MRI imaging. Another type of enzyme whose levels are known to be upregulated in various cancer subtypes is β-galactosidase (β-gal). Sharma et al. have developed theragnostic prodrug for the treatment of colon cancer using receptor-mediated targeting and enzyme responsive activation. In this study, β-gal was used for both targeting asialoglycoprotein (ASGP) receptors and activation of prodrug.

## 6.9 NANOPARTICLES ACTIVATION BY REDOX

The intracellular GSH levels inside TME are in the range of $0.5–10 \times 10{-}3$ M, which is four, times higher than the GSH levels in normal tissues. Intracellular compartments such as the cytosol, mitochondria, and cell nucleus are known to contain a much higher concentration of GSH than extracellular fluids. Such drastic differences in the GSH level between TME and other normal tissue could be utilized as a promising platform to design nanoparticles to selectively release therapeutic drugs in a triggered fashion after delivery to the tumor cells. The introduction of bio-reducible disulfide bonds has attracted much interest in the design of redox-responsive nanoparticles that can release their payloads efficiently in intracellular reductive environments. Sun et al. have reported the synthesis of a redox-sensitive drug delivery system for the treatment of laryngopharyngeal carcinoma. The redox-sensitive amphiphilic polymer was synthesized by conjugating heparosan with deoxycholic acid through disulfide bonding. The polymer formed self-assembled nanoparticles that can disassemble via reductive cleavage of the disulfide bonds and trigger drug release in the intracellular environment. Our group has also reported the synthesis of zwitterionic polymer-based hybrid nanoparticles with glutathione (GSH) and endosomal pH-responsiveness. GSH-responsive drug delivery systems could selectively deliver the drug in TME and enhance the antitumor efficacy of the nanoparticle. Zhou et al. have reported the synthesis of redox-sensitive drug delivery systems based on dextran and indomethacin. Redox responsive polymer (DEX-SS-IND) was fabricated through the introduction of a disulfide bridge (cystamine) in between based on dextran and indomethacin. The anti-cancer drug, DOX was encapsulated inside the core-shelled micelles formed by self-assembly of DEX-SS-IND. In reducing the environment, the DEX-SS-IND depolymerizes and releases DOX. *In-vivo* Schematic illustration of self-assembled micelle and GSH triggered the release of DOX. Xia et al. have reported the synthesis of polycarbonate-based core-crosslinked redox responsive nanoparticles (CC-RRNs) for the targeted delivery of DOX. CC-RRNs were synthesized by the click reaction between PEG-b-poly(MPC)n (PMPC), α-lipoic acid and 6-bromohexanoic acid. The di-sulfide cross-linked core is formed by the addition of a catalytic amount of dithiothreitol (DTT). CC-RRNs demonstrated a controlled release of DOX under redox condition. Such multifunctional responsive systems hold the key for future developments in TME-assisted nanomedicine. However, it would be noted that exact intracellular fate of redox-sensitive nanoparticles is not clearly understood. Studies have reported that cell surface thiols can affect the internalization of di-sulfide conjugated peptides. Hence, a better understanding about the intracellular trafficking of the nanoparticles is required for the development of nanoparticle-activated by redox environment. This higher level of ROS in tumors could be utilized for the development of ROS-responsive nanoparticles, which could enhance site-specific drug release. The most commonly used characteristic groups employed for the development of ROS-responsive systems are boronic ester, thioketal, and sulfide groups. Such ROS-responsive systems can lead to the development of drug carriers for efficient delivery of chemotherapy. Sun et al. have developed ROS-responsive micelles for enhanced drug delivery applications. For the development of ROS-responsive micelles, a ROS-sensitive thioketal linker with a π-conjugated structure was conjugated into methoxy (polyethylene glycol) thioketal-poly(ε-caprolactone) (mPEG-TK-PCL) micelles. The micelles were formed through the self-assembly of mPEG-TK-PCL and

DOX was then loaded through physical encapsulation. The DOX-loaded mPEG-TK-PCL micelles demonstrated enhanced anticancer activity due to the rapid cleavage of the thioketal linker in the presence of increased ROS levels in cancer cells, thereby accelerating drug release and augmenting cancer cell inhibition. The prodrug was then used as a drug carrier to further encapsulate DOX and form DOX-loaded prodrug micelles. DOX-loaded prodrug micelles demonstrated superior anti-tumor efficacy over non-responsive DOX-loaded poly(ethylene glycol)-block-polycaprolactone (PEG2k-PCL5k) micelles. Yu et al. have reported the synthesis of chalcogen containing polycarbonate for ROS responsive PDT. The ROS responsive polycarbonate was prepared by the ring-opening polymerization of cyclic carbonate monomers with ethyl selenide, phenyl selenide or ethyl telluride group. PEG was employed as a macro-initiator to prepare amphiphilic block co-polymers, which form spherical nanoparticles of less than 100 nm. These nanoparticles completely dissociate in the presence of ROS while remain stable in neutral phosphate buffer. To check the ROS responsive drug release potential of these nanoparticles, DOX, and Ce6 were loaded. Upon laser irradiation, Ce6 would generate $O_2$, which will trigger the degradation of the nanoparticles resulting in the faster release of DOX. Even though numerous ROS responsive nanoparticles have been reported for biomedical application, there are several challenges needed to be addressed such as the biocompatibility of the ROS sensitive linker used, the stability of the linker during circulation and at the normal cells. Since the levels of ROS changes with variations in patients and disease conditions the selection of linkers and carriers should be intensively considered for personalized application [31, 32].

## KEYWORDS

- catalase
- dendritic cells
- hyaluronic acid
- immune stimulatory peptides
- low-density lipoproteins
- matrix metalloproteinase

# REFERENCES

1. Merriam-Webster's Collegiate Dictionary, (2003). *Springfield, Mass* (11th edn.). Merriam-Webster, Inc.
2. Kaufmann, M., Von Minckwitz, G., Bear, H. D., Buzdar, A., McGale, P., Bonnefoi, H., et al., (2007). Recommendations from an international expert panel on the use of neoadjuvant (primary) systemic treatment of operable breast cancer: New perspectives-2006. *Ann. Oncol., 18*(12), 1921–1934.
3. Dawood, S., Merajver, S. D., Viens, P., Vermeulen, P. B., Swain, S. M., Buchholz, T. A., et al., (2011). International expert panel on inflammatory breast cancer: consensus statement for standardized diagnosis and treatment. *Ann. Oncol., 22*(3), 511–523.
4. Rakha, E. A., Reis, J. S., & Ellis, I. O., (2010). Combinatorial biomarker expression in breast cancer. *Breast Cancer Res. Treat, 120*(2), 291–308.
5. De Milito, A., & Fais, S., (2005). Tumor acidity, chemoresistance and proton pump inhibitors. *Future Oncol., 1*(6), 771–786.
6. Gottesman, M. M., & Pastan, I., (1993). Biochemistry of multidrug-resistance mediated by the multidrug transporter. *Annu. Rev. Biochem., 62*, 381–427.
7. Cole, S. P. C., Bhardwaj, G., Gerlach, J. H., Mackie, J. E., Grant, C. E., Almquist, K. C., et al., (1992). Overexpression of a transporter gene in a multidrug-resistant human lung-cancer cell-line. *Science, 258*(5088), 1651–1654.
8. Langer, R., (1998). Drug delivery and targeting. *Nature, 392*(6679), 1–10.
9. Ren, F., Chen, R., Wang, Y., Sun, Y., Jiang, Y., & Li, G., (2011). Paclitaxel-loaded poly(n-butylcyanoacrylate) nanoparticle delivery system to overcome multidrug resistance in ovarian cancer. *Pharmaceutical Research (Dordrecht), 28*(4), 891–906.
10. Niederhausern, S., Bondi, M., Messi, P., Iseppi, R., Sabia, C., Manicardi, G., et al., (2011). Vancomycin-resistance transferability from van A enterococci to staphylococcus aureus. *Curr. Microbiol., 62*(5), 1361–1367.
11. Raghunath, D., (2008). Emerging antibiotic resistance in bacteria with special reference to India. *J. Biosci., 33*(4), 591–603.
12. Werner, G., Strommenger, B., & Witte, W., (2008). Acquired vancomycin resistance in clinically relevant pathogens. *Future Microbiol., 3*(5), 541–562.
13. Yoshikawa, T., Okada, N., Oda, A., Matsuo, K., Matsuo, K., Mukai, Y., et al., (2008). Development of amphiphilic gamma-PGA-nanoparticle based tumor vaccine: potential of the nanoparticulate cytosolic protein delivery carrier. *Biochemical and Biophysical Research Communications, 366*(2), 401–413.
14. Zaks, K., Jordan, M., Guth, A., Sellins, K., Kedl, R., Izzo, A., et al., (2006). Efficient immunization and cross-priming by vaccine adjuvants containing TLR3 or TLR9 agonists complexed to cationic liposomes. *Journal of Immunology, 176*(12), 7331–7345.
15. Janeway, C. A., Travers, P., Walport, M., & Schlomchick, M. J., (2001). *Immunobiology* (5th edn.). New York: Garland Science.
16. Banchereau, J., & Steinman, R. M., (1998). Dendritic cells and the control of immunity. *Nature, 392*(6673), 241–252.
17. Steinman, R. M., (2003). The control of immunity and tolerance by dendritic cell. *Pathol. Biol. (Paris), 51*(2), 51–60.
18. Lemoine, D., & Preat, V., (1998). Polymeric nanoparticles as delivery system for influenza virus glycoproteins. *J. Control Release, 54*(1), 11–27.

19. Uto, T., Wang, X., Sato, K., Haraguchi, M., Akagi, T., Akashi, M., et al., (2007). Targeting of antigen to dendritic cells with poly(gamma-glutamic acid) nanoparticles induces antigen-specific humoral and cellular immunity. *Journal of Immunology, 178*(5), 2971–2986.

20. He, C., Hu, Y., Yin, L., Tang, C., & Yin, C., (2010). Effects of particle size and surface charge on cellular uptake and biodistribution of polymeric nanoparticles. *Biomaterials, 31*(13), 3651–3666.

21. Heit, A., Schmitz, F., Haas, T., Busch, D. H., & Wagner, H., (2007). Antigen co-encapsulated with adjuvants efficiently drive protective T-cell immunity. *Eur. J. Immunol., 37*(8), 2061–2074.

22. Fang, R. H., Aryal, S., Hu, C. M. J., & Zhang, L. F., (2010). Quick synthesis of lipid-polymer hybrid nanoparticles with low polydispersity using a single-step sonication method. *Langmuir, 26*(22), 16951–16962.

23. Audran, R., Peter, K., Dannull, J., Men, Y., Scandella, E., Groettrup, M., et al., (2003). Encapsulation of peptides in biodegradable microspheres prolongs their MHC class-I presentation by dendritic cells and macrophages *in vitro*. *Vaccine, 21*(11/12), 1251–1255.

24. De Jong, S., Chikh, G., Sekirov, L., Raney, S., Semple, S., Klimuk, S., et al., (2007). Encapsulation in liposomal nanoparticles enhances the immunostimulatory, adjuvant, and anti-tumor activity of subcutaneously administered CpG ODN. *Cancer Immunology, Immunotherapy, 56*(8), 1251–1264.

25. Diwan, M., Tafaghodi, M., & Samuel, J., (2002). Enhancement of immune responses by co-delivery of a CpG oligodeoxynucleotide and tetanus toxoid in biodegradable nanospheres. *J. Control Release, 85*(1–3), 241–262.

26. Jiang, W. L., & Schwendeman, S. P., (2008). Stabilization of tetanus toxoid encapsulated in PLGA microspheres. *Molecular Pharmaceutics, 5*(5), 801–817.

27. Fournier, P., & Schirrmacher, V., (2009). Randomized clinical studies of anti-tumor vaccination: state of the art in 2008. *Expert Rev. Vaccines, 8*(1), 51–66.

28. Gattinoni, L., Powell, D. J., Jr., Rosenberg, S. A., & Restifo, N. P., (2006). Adoptive immunotherapy for cancer: Building on success. *Nat. Rev. Immunol., 6*(5), 381–393. PMCID: 1473162.

29. Di Lisi, D., Bonura, F., Macaione, F., Peritore, A., Meschisi, M., Cuttitta, F., et al., (2011). Chemotherapy-induced cardiotoxicity: Role of the tissue Doppler in the early diagnosis of left ventricular dysfunction. *Anti-Cancer Drugs, 22*(5), 461–472.

30. Hong, R. L., Huang, C. J., Tseng, Y. L., Pang, V. F., Chen, S. T., Liu, J. J., et al., (1999). Direct comparison of liposomal doxorubicin with or without polyethylene glycol coating in C-26 tumor-bearing mice: Is surface coating with polyethylene glycol beneficial? *Clin. Cancer Res., 5*(11), 3641–3652.

31. Christiansen, S., (2011). Clinical management of doxorubicin-induced heart failure. *Journal of Cardiovascular Surgery, 52*(1), 131–137.

32. Hydock, D. S., Lien, C. Y., Jensen, B. T., Schneider, C. M., & Hayward, R., (2011). Exercise preconditioning provides long-term protection against early chronic doxorubicin cardiotoxicity. *Integrative Cancer Therapies, 10*(1), 41–57.

# CHAPTER 7

# Bioresponsive Hydrogels for Controlled Drug Delivery

TAMGUE SERGES WILLIAM, DIPALI TALELE, and DEEPA H. PATEL

*Department of Pharmaceutics, Parul Institute of Pharmacy and Research, Faculty of Pharmacy, Parul University, P.O. Limda, Ta. Waghodia, Vadodara–391760, Gujarat, India, Phone: 02668-260287, Fax: 02668-260201, E-mails: deepaben.patel@paruluniversity.ac.in, Pateldeepa18@yahoo.com (D. H. Patel)*

## 7.1 INTRODUCTION OF HYDROGELS

Hydrogels can be defined as polymeric matrices consists of a cross-linked three-dimensional networks with interstitial spaces that contain as much as 90–99% w/w of water, and are capable also to retain significant amount of water in their structure (from 10 to 20% up to thousands of times their dry weight), without dissolving in water, it is called swelling.

Hydrophilic gels or hydrogels derive their ability to absorb water from hydrophilic functional groups attached to their polymeric backbone, while their resistance to dissolution arises from crosslinks between network chains.

The term hydrogel appeared for the first time in an article published in 1894 according to Lee, Kwon, and Park, even if the material described was not really hydrogel as described today. It was indeed colloidal gel-based inorganic salt. The first crosslinked network material complying with all the characteristics of typical hydrogel was reported by Wichterle and Lim named pHEMA (polyhydroxyethylmethacrylate), which had the goal to be used in permanent contact with human tissues. According to Kopecek [2], hydrogels were the first biomaterials developed for human uses [1, 2].

Buwalda et al. in 2014 [3] have suggested that history of hydrogel could be resumed in three blocks corresponding to different generations [3, 4]:

1. The first generation of hydrogel, two decades after the discovery of pHEMA was essentially focused on a wide range of crosslinking procedures involving polymerization of a water-soluble monomer in presence of crosslinker or crosslinking of natural polymers or water-soluble synthetic polymers with an initiator. The general aim was to develop a material with high swelling, good mechanical properties, and relatively simple rationale.

2. In the 1970s, studies converged towards the novel type of materials that were capable to react to changes in environmental conditions (specific stimuli) such as variation in temperature, pH, or concentration of biomolecules, and respond by structural modification of hydrogel such as, swelling, gel formation, degradation, and drug release. Hydrogels was mainly cross-linked via hydrophobic and ionic interactions.

3. The third generation was characterized by the advent of novel and diverse types of physical interactions exploited as crosslinking techniques such as metal-ligand coordination, stereocomplexation, peptide interactions, and inclusion complex formation.

Over past decades, there has been a lot of progress in hydrogel development, research quickly moved from a simple network to 'smarts hydrogels' with a wide range of tunable properties and trigger stimuli, aimed to engage in the biological environment.

The hydrogels development have gained significant interest in biomedical application like biological sensing, drug delivery, and tissue regeneration because of their advantages such as their structural similarity to the natural extracellular matrix (ECM), their highly tunable mechanical properties, their high permeability, their potential biocompatibility, and biodegradability. This chapter will highlight recent developments, synthesis, characterization, and application of hydrogels with biological responsiveness built-in.

## 7.2 CLASSIFICATION OF HYDROGELS

Hydrogels can be categorized on the basis of their origin, polymeric composition, the type of crosslinking and physical appearances.

## 7.2.1 CLASSIFICATION BASED ON THEIR ORIGIN

On the basis of their source or origin, it is possible to distinguish three mains categories of hydrogels: natural, synthetic, and hybrid hydrogels [4, 5].

### 7.2.1.1 NATURAL HYDROGELS

Natural hydrogels consist of natural bio-macromolecules including proteins and polysaccharides (Figure 7.1). These bio-macromolecules can be obtained from different origins like animals, plants, or microorganisms. Most of them are natural constituents of extracellular matrices.

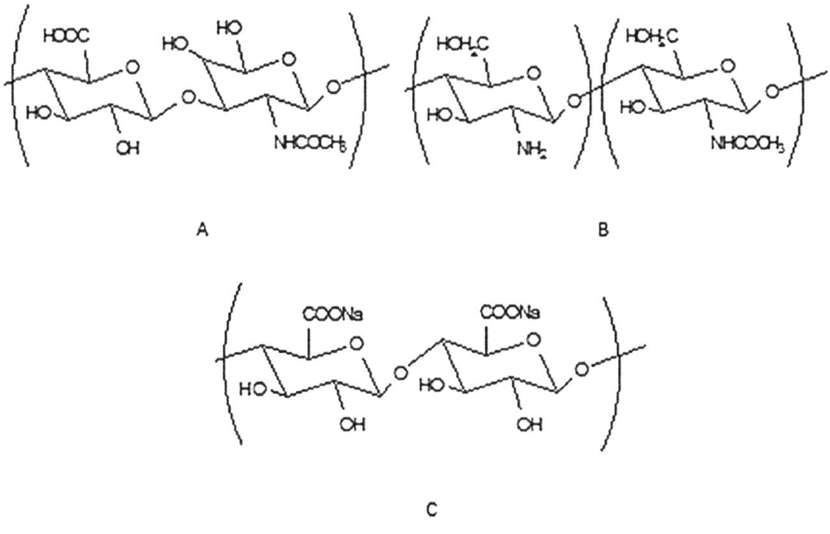

**FIGURE 7.1**   Chemical structures of (A) hyaluronic acid, (B) chitosan, and (C) sodium alginate.

Natural hydrogels are generally prepared using polysaccharides such as hyaluronic acid (HA), chitosan (CS), alginate, cellulose, starch or protein, including collagen, fibrin. A number of polysaccharide-based or protein-based hydrogels have been developed which are used in the application of cartilage and bone tissue engineering. Mostly hydrogels have been prepared using HA, CS, and alginate biopolymers which are established examples of polysaccharides in tissue engineering [5].

HA is an abundant non-sulfated glycosaminoglycan (GAG) occur-ring naturally in all living organisms. It is a component of synovial fluid and extracellular matrices composed of repeating disaccharide units of d-glucuronic acid and N-acetyl-d-glucosamine linked by $\beta$-1-3 and $\beta$-1-4 glycosidic bonds, [(1→3)-$\beta$-dGlcNAc-(1→4)-$\beta$-d-GlcA-] [6].

CS is a $\beta$-(1-4) linked polysaccharide of D-glucosamine produced by the deacetylation of chitin. It is a biodegradable, biocompatible, and non-toxic natural polymer derived from natural sources, which is the exoskeleton of insects and crustacean [7].

Alginates are linear unbranched copolymers that consist of homopoly-meric blocks of (1,4)-linked b-D-mannuronic acid (M) and its C-5 epimer, a-L-glucoronic acid (G) residues. They are isolated from brown algae such as *Laminaria hyperborea* and lessonia found in coastal water around the globe [8].

Collagen-based hydrogels have been developed and widely used due to its ubiquitous presence in different tissues of body. While gelatin obtained from hydrolysis of collagen is significantly less expensive. Several cross-linking strategies were performed to control mechanical properties and 3D structures using, for example, 1,4-butanediol diglycidylether (BDDGE) or genipin cross-linking.

Natural polysaccharides based hydrogels have been produced by various methods including physical treatment such as UV-assisted photo-polymerization, freeze-drying of CS; chemical or covalent cross-linking by esterification, carbodiimide chemistry (HA, alginates), using a crosslinking agent such as genipin, glutaraldehyde, squarate adipic hydrazide (HA, CS) [5].

Natural polysaccharides based hydrogels are of great interest for biomaterials scientists because they provide intrinsic bioactivities such as cell adhesion, biodegradability. They are biocompatible, possess a lower toxicity which is generally due to crosslinking agents, they can actively support cell viability or cell differentiation and have low-cost production. However, natural hydrogels present some limitations that restrict their use:

- Potential immunogenicity;
- Poor understanding of their mechanical properties and their depen-dence on polymerization or gelation conditions;
- Batch to batch variations is due to their natural origin like bovine fibrinogen, rat tail collagen.

To control the final microstructures and properties was found to be difficult due to the variations in reproducibility.

## 7.2.1.2 SYNTHETIC HYDROGELS

Synthetic materials based hydrogels have been designed to overcome the limitations observed with natural hydrogels. Synthetic hydrogels are more reproducible, easily synthesized in large quantities with controlled parameters such as molecular weight, molecular architectures, and microscopic morphologies. Using polymers with controlled molecular weight and biodegradable linkers allows for the fine-tuning of chemical composition, mechanical properties, and degradation rates of hydrogel.

Hydrolytically stable cross-linked poly(2-hydroxyethyl methacrylate) HEMA is a poly(acrylic acid) (PAA) derivative that has been used in various applications such as drug delivery, also for making contact lens in ophthalmic purposes. Many different types of molecules and cells have been encapsulated into poly(HEMA) gels, and had shown successful results for delivery of insulin and other proteins. Nondegradability of poly(HEMA) gels in physiological conditions limits its use [25].

Poly-*N*-isopropylacrylamide (pNIPAM) is a thermo-responsive polymer with inverse solubility and a reversible phase transition upon heating. It has a LCST value of approximately 32°C where below 32°C, polymer is hydrophilic and water soluble and above 32°C, it is hydrophobic and which then becomes a viscous gel strongly adhere to tissue. pNIPAM is soluble at room temperature but its phase separates at body temperature (37°C); thus it can be used as a linear polymer, a hydrogel, or a copolymer. pNIPAM polymer has been used in thermo-sensitive coatings or micelles for controlled release of the drug, drug targeting in solid tumors with local hyperthermia, in eye drop preparations and as a new embolic material in neurosurgery [26].

Polyethylene glycol (PEG) is one of the most frequently applied synthetic hydrophilic polymers for hydrogel preparation and currently approved by the Food and Drug Administration (FDA) for several biomedical applications. Various non-immunogenic synthetic materials have been tested as 3D hydrogel scaffolds for cell transplant applications, including PEG, polyvinyl alcohol (PVA), PAA, and various polypeptides. These hydrogels can be polymerized *in situ* under mild physiological conditions [4] (Table 7.1).

**TABLE 7.1**  Some Example of Naturals Polymers and Synthetic Monomers Used in Hydrogels Preparation

| Natural Polymers | Synthetic Monomers/Polymers |
|---|---|
| Hyaluronic acid | Hydroxyethyl methacrylate (HEMA) |
| Chitosan | Vinyl acetate (V Ac) |
| Alginate | Polyethylene glycol (PEG) |
| Collagen | N-(2-Hydroxy propyl) methacrylate (HPMA) |
| Gelatin | N-Vinyl-2-pyrrolidone (NVP) |
| Fibrin | N-Isopropylacrylamide (NIPAMM) |

## 7.2.1.3   HYBRID HYDROGELS

According to Jia and Kick [9], hydrogel systems composed of chemically, morphologically, and functionally different building blocks interconnected via chemical or physical means [9].

Although natural and synthetic hydrogels have shown promising application in tissue repairs, hybrid hydrogels development was exploited in tissue engineering with the goal to respond to critical issues observed with traditional hydrogels regarding their bioactivities, mechanical strength, and degradation kinetics. Traditional hydrogels are heterogeneous bulk gels made up of inert synthetic polymers randomly interconnected, with network defects such as entanglements, chain ends, and phase-separated regions. The uncontrolled heterogeneity at different length scales dramatically affects the diffusion characteristics, mechanical properties, and cellular functions [9].

Many traditional hydrogels do not exhibit biological activities, hierarchical organization, and structural integrity that are necessary to facilitate cell infiltration and neovascularization. Hybrid hydrogels combine the advantages of naturally derived hydrogel and synthetic hydrogel. It combines the strengths of intrinsic bioactivity from naturally derived materials and superior control over network physical and chemical properties from synthetic materials (Figure 7.2) [9, 10].

The most common approach in three-dimensional (3D) hybrid matrices is the use of matrix metalloproteinase (MMP)-sensitive peptides derived from native ECM molecules to crosslink synthetic polymers [10].

Some examples of hybrid hydrogels include collagen mimetic hybrid hydrogel and elastin mimetic hybrid hydrogel:

- A (2)-methacryloyloxyethylphosphorylcholine (MPC) immobilized collagen that has been developed by Nam et al. [11]. They used 1-ethyl-3-(3-dimethyl aminopropyl)-1-carbodiimide hydrochloride (EDC) and N-hydroxysuccinimide (NHS) to cross-link a collagen film in 2-morpholinoethane sulfonic acid (MES) buffer. To cross-link MPC to collagen gel, poly(MPC-co-methacrylic acid) (PMA) having a carboxyl group side chain was chosen [11].
- Grieshaber et al. [12] have synthesized by step-growth polymerization, a multiblock elastin-mimetic hybrid polymers (EMHPs) containing flexible synthetic segments based on PEG alternating with alanine-rich, lysine-containing peptides [12].

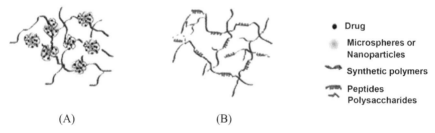

(A)  (B)

**FIGURE 7.2**   Schematic representation of hybrid hydrogel; (A) microscopic level hybridization approaches, (B) molecular level hybridization approaches.

## 7.2.2   CLASSIFICATION BASED ON POLYMERIC COMPOSITION

The method of preparation will determine the polymeric constitution of hydrogels. Hydrogels can be classified into three groups:

1. **Homopolymeric Hydrogels:** It is a polymer network consisting of a single species of monomer, which is the basic structural unit comprising of any polymer network. Homopolymers may have cross-linked skeletal structure dependent on the nature of the monomer and polymerization method [13].
2. **Co-Polymeric Hydrogels:** It is made up of two or more distinct monomer species with at least one hydrophilic component, assembled in a random, block or alternating configuration along the chain of polymer [14].

**3. Multi-Polymer or Interpenetrating Polymeric Hydrogel (IPN):**
It is made of two independent cross-linked synthetic and/or natural polymer components, confined in a network form [15].

### 7.2.3 CLASSIFICATION BASED ON CONFIGURATION

This classification of hydrogels relies on their physical structure and chemical composition:

1. amorphous (non-crystalline);
2. semi-crystalline: a complex mixture of amorphous and crystalline phases;
3. crystalline.

### 7.2.4 CLASSIFICATION BASED ON TYPE OF CROSSLINKING

On the basis of their chemical or physical behavior of the cross-link junctions, two types are distinguished:

- Chemically cross-linked networks have stable junctions, while physical cross-linked networks have temporary junctions which are the results of either polymer chain entanglements or physical interactions such as ionic interactions, hydrogen bonds or hydrophobic interactions [16].

### 7.2.5 CLASSIFICATION BASED ON PHYSICAL APPEARANCE

Depending on the polymerization method employed during the formulation process, hydrogels can appear as matrix, film, or microspheres.

### 7.2.6 CLASSIFICATION BASED ON THE CHARGE OF THE CROSSLINKED CHAIN

Based on the presence or absence of electrical charge on the crosslinked chains, four groups can be distinguished:

1. nonionic (neutral);
2. ionic (anionic or cationic);
3. ampholytic (comprising both acidic and basic groups); and
4. zwitterionic (polybetaines) consisting of both anionic and cationic groups in each structural repeating unit.

## 7.3 PHYSICAL AND CHEMICAL PROPERTIES OF HYDROGELS

### 7.3.1 SWELLING PROPERTIES

Hydrogels are cross-linked polymer networks that are able to swell in a liquid medium. Hydrogels may absorb from 10–20% up to thousands of times of their dry weight in water. When a dry hydrogel starts to absorb water, the initial water molecules moving into the matrix will hydrate the most polar, hydrophilic groups, leading to primary bound water. This leads to the swelling of hydrogel linkage and exposes hydrophobic groups which also intermingle with water molecules, resulting in hydrophobically-bound water or secondary bound water. Primary and secondary bound water are often merged and solely called as total bound water. Progressively as the network swells, if the network chains are degradable, the gel will initiate to disintegrate and dissolve at a rate depending on its composition. Various methods are employed to evaluate the relative amounts of bound and free water contained in the hydrogel, such as differential scanning calorimetry (DSC) and nuclear magnetic resonance (NMR) [17].

The swelling is the property to absorb water and retain it for a relative long time. It can be estimated by measuring the dry weight and the swollen state weight and calculating the swelling ratio by using the formula

$$WU = \text{swollen weight} - \text{dry weight/dry weight} \times 100$$

According to Griffith [18], the assessment of swelling is the crucial assay to determine the crosslinking degree, mechanical properties, and degradation rate. For numerous gels, the estimation of swelling and swollen state stability is an easy, cheapest, and assured way to differentiate between crosslinked gels and the non-crosslinked original polymer [18].

### 7.3.2 POROSITY

Porosity is a morphological characteristic of a material that can be illustrated as the presence of a void cavity inside the bulk. Pores may be created in hydrogels by the process of phase separation in the course of synthesis or they may present as smaller pores within the network. In a sample, pores can display distinct morphologies, i.e., they can be closed, open as a blind end, or interconnected, again divided into cavities and throats.

Pore-size distributions of hydrogels are influenced by three factors:

1. Concentration of the chemical cross-links of the polymer strands;
2. Concentration of the physical entanglements of the polymer strands; and
3. Net charge of the polyelectrolyte hydrogel.

The porous structure of a hydrogel is also influenced by the properties of the surrounding solution, principally by dissolved ionic solutes (Donnan effects) and by dissolved uncharged solutes which separate unevenly between the gel phase and the solution phase (Osmotic effects) [17].

It is somehow difficult to evaluate the porosity of hydrogel due to some inconvenient parameters such as temperature change, mathematical manipulation, and postulation required. Porosity is an important parameter that needs to be controlled in many devices for various applications such as optimal cell migration in hydrogel-based scaffolds or tunable lode/ release of macromolecules [17].

$$\text{Porosity } \% = V_{pore}/V_{bulk} + V_{pore} \times 100$$

Many techniques have been employed in the past decades for porosity investigation studies. First of all, porosity can be estimated by theoretic procedures like unit cube analysis, mass technique, Archimedes method, liquid displacement method, these evaluations are generally coupled with optical and electronic microscopy. Other methods include the mercury porosimetry based on Washburn's equation, gas pycnometry, gas adsorption (small quantity adsorption, monolayer, and multilayer adsorption), liquid extrusion porosity, and capillary flow porosity. Moreover, an alternate assay is the Micro-CT also called X-ray Microtomography, a relative new imaging technology expressed as nondestructive high-resolution radiography, qualified for qualitative and quantitative assays on samples and estimation of their pore interconnections. Micro-CT can provide information on average pore size, pore size distribution, and pore interconnection which are essential factors of a hydrogel matrix that are often difficult to calculate. Micro-CT provides also information about wall thickness and anisotropy/ isotropy of the sample [19].

### 7.3.3 MECHANICAL PROPERTIES

The evaluation of a mechanical property is of great importance in various biomedical applications such as matrix for drug delivery, ligament, and tendon repair, wound dressing material, tissue engineering, and cartilage replacement material. The mechanical properties of hydrogel should be suitable enough to allow less modification of its physical texture during the delivery of the therapeutic agent for the predetermined period of time.

Upon a change in its degree of crosslinking the desired mechanical properties of the hydrogel are achieved [20]. So by increasing the cross-linking degree, a stronger hydrogel is obtained while by lowering it a weak structure is obtained. The mechanical properties of hydrogel can be assessed by a dynamic mechanical analysis (DMA) rheometer [21].

The crosslinking degree can be correlated to basically with every characteristic of a hydrogel. The nature of the crosslinking can vary a lot and can be distinguish two main categories of crosslinking.

Physical crosslinking: For example, hydrophobic interactions between chains, ionic interactions between a polyanion and a polycation (complex coacervation), or ionic interactions between a polyanion and multivalent cations (ionotropic hydrogel).

Chemical crosslinking: By ultraviolet (UV) irradiation, heating or chemical crosslinking via crosslinker with a huge ensemble of reactions, such as Michael's reaction, Michaelis-Arbuzov reaction, and nucleophile addition [31].

### 7.3.4 BIOCOMPATIBLE PROPERTIES

Hydrogels fulfilled two important criteria which are biocompatibility and safety (non-toxic) gained great interest and pertinent in the biomedical field. Polymers must pass cytotoxicity and *in vivo* toxicity tests. Biocompatibility is the capability of a material to function with an appropriate host response in a specific application. Biocompatibility consists basically of two parameters namely biosafety and bio-functionality

1. **Biosafety:** It is the adequate host response not only systemic but also local (i.e., surrounding tissue), the absence of cytotoxicity, mutagenesis, and carcinogenesis.
2. **Bio-Functionality:** It is the capacity of a material to perform the specific task for which it is intended. This explanation is exceptionally applicable in tissue engineering since the nature of tissue construct is to constantly interact with the body through the healing and cellular regeneration process as well as during scaffold degradation. Moreover, initiators, organic solvents, stabilizers, emulsifiers, unreacted monomers, and crosslinkers utilized in polymerization and hydrogel synthesis may be toxic to host cells if they ooze out to tissues or encapsulated cells. To eradicate harmful chemicals from preformed gels, certain purification processes should be implemented such as solvent washing or dialysis [17, 20].

## 7.4 BIO-RESPONSIVE HYDROGELS

Bio-responsive or biologically responsive hydrogels are dynamic systems that are capable to sense and respond to signal/biological events by changing their properties. They are smart materials designed to react to the recognition of a biological agent or to the variation of a specific parameter of their biological environment, which are considered as their triggers.

The biological agent may be biomolecules (sugars, proteins, nucleic acid, and enzymes), small molecules such as metabolites and peptides or even whole cells such as endothelial cells. Other stimuli include pH variation, temperature, ionic strength, solvent polarity, light, and electric/ magnetic field [23].

The interactions between hydrogels and the triggers lead to various type of response which includes among others swelling, collapsing, degradation, and release, mechanical deformation, solution to gel transition, optical density variation, or electronic variations.

The response may be binary for example in presence or absence of the biological agent at a certain limit. The response may also be scalable with the chemical potential or activity of the biological agent [23].

## 7.5 PREPARATION OF BIO-RESPONSIVE HYDROGELS

In general, the three parts of hydrogels preparation are, monomer, initiator, cross-linker. Hydrogel can be obtained by copolymerization or crosslinking free radical polymerization of hydrophilic monomers with multifunctional cross-linkers. Or mostly water-soluble natural or synthetic linear polymers are cross-linked to form hydrogel. The crosslinking techniques are described below:

1. Physical crosslinking (entanglements, electrostatic, and crystallite formation);
2. Chemical crosslinking, using chemical reactions to link polymers;
3. Radiation crosslinking (using ionizing radiation to generate main chain free radicals).

Modification improves the mechanical properties and viscoelasticity for use in biomedical and pharmaceutical fields.

### 7.5.1 PHYSICAL CROSSLINKING

It is the crosslinking of hydrogel without using any chemical additives. It is the most employed route due to the relative ease of production and the advantage of the absence of cross-linkers used for synthesis, where the polymer is cross-linked using physical interactions. These techniques involve interactions of ions such as hydrogen bonding, polyelectrolyte complexation, and hydrophobic association. Physical crosslinking methods include:

1. **Heating/Cooling a Polymer Solution:** Cooled hot solution of gelatin and carrageenan leads to formation of physically cross-linked hydrogels. The gel is formed due to helix-formation, association of helices and forming junction zones. Carrageenan is present as random coil conformation when it is in hot solution above the melting transition temperature but on cooling, it transforms to rigid helical rods. Double helices further aggregate to form a stable gel in presence of salt due to the screening of repulsion of the sulfonic group. Hydrogels can be often obtained by simply warming the polymer solution that induces the block

copolymerization. Polyethylene glycol-polylactic acid hydrogel and Polyethylene oxide-polypropylene oxide are its examples.

2. **Complex Coacervation:** Complex coacervate gels are the results of mixing a polyanion with a polycation. The principle behind this method is the attraction of polymers with opposite charges which will stick together and form soluble and insoluble complexes depending on parameters like concentration and pH of the respective solutions. Below its isoelectric point, proteins are positively charged and likely to associate with anionic hydrocolloids and form polyion complex coacervate. The perfect illustration of this method is the coacervation of polycationic CS with polyanionic xanthan (Figure 7.3).

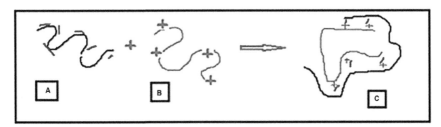

**FIGURE 7.3**   Complex coacervation (C) between polyanion (A) and polycation (B).

3. **Ionic Interaction:** Crosslinking between polymers may occur also due to the addition of di- or trivalent counterions in the ionic polymer. This method leans on the principle of gelling polyelectrolyte solution ($Na^+$, alginate-) with a multivalent ion of opposite charges (e.g., $Ca^{2+}$ +2Cl$^-$). Other examples are CS poly-lysine, CS dextran hydrogels [24].

4. **Hydrogen Bonding:** Hydrogen bond can be formed by the association of an electron-deficient Hydrogen atom and a high electron density functional group. Hydrogen bonded hydrogel also can be obtained by lowering the pH of an aqueous solution of polymers carrying carboxyl groups. Hydrogen bound carboxymethyl cellulose (CMC) network formed by dispersing CMC into 0.1M HCl. Another example of hydrogel obtained by lowering the pH to form H-bonded hydrogel is PAA and polyethylene oxide (PEO-PAAc) based hydrogel.

5. **Maturation:** Also called heat-induced aggregation, is a method of hydrogel formation which is based on the aggregation of proteinous components of natural polymers. Aggregation occurs by applying a heat treatment that increases the molecular weight and subsequently produces a hydrogel form with enhanced mechanical properties and water binding capability. It has been reported that with this method, hydrogel can be produced with precisely structured molecular dimension. This method is mostly applied with gums such as, gum Arabic, gum ghatti, Acacia karensis. Gum Arabic is mainly constituted of carbohydrates but contains 2–3% of proteins as an integral part of its structure.

6. **Freeze Thawing:** A freeze-thaw cycle is a reputed method to achieve physical crosslinking of polymer to form hydrogel. The technique is based on the formation of microcrystals in the structure due to freeze-thawing. Gel of polyvinyl alcohol and xanthan is an example of gel obtained by this method.

7. **From Amphiphilic Graft and Block Polymer:** Based on the ability of graft and block polymer to self-assemble in aqueous media to form hydrogels and polymeric micelles in which the polymers hydrophobic parts are self-assembled. Multi-block polymers made up of hydrophobic chains which have hydrophilic grafts or a water-soluble polymer backbone to which hydrophobic segments are attached. PEG-poly(γ-benzyl L-glutamate) and thermo-sensitive hydrogels from PEG-PNIPAM is an example of a multi-block polymer.

### 7.5.2 CHEMICAL CROSSLINKING

Chemical crosslinking is obtained by using a crosslinking agent to link two polymer chains and grafting of monomers on the backbone of the polymers. Natural and synthetic polymer can be crosslinked through their functional groups such as OH, COOH, and $NH_2$ with cross-linkers such as aldehyde (e.g., glutaraldehyde and adipic acid dihydrazide). IPN is a polymerized monomer within another solid polymer to form interpenetrating network structure and hydrophobic interactions are other methods to obtain chemically permanent crosslinking.

1. **Crosslinking with Aldehydes:** Hydrophilic polymers consist of -OH groups may be cross-linked using aldehydes as crosslinking

agents. The reaction conditions are low pH, high temperature, and methanol added as quencher. Polymers having amines groups can be cross-linked through the same process but at mild conditions.

2. **Crosslinking by Addition Reaction and Condensation Reaction:** Addition reaction may be used to crosslink functional groups of hydrophilic polymers with higher functional cross-linkers. Polysaccharide which may be cross-linked by means of divinylsulfone is an example of addition reaction. Hydrogels synthesis can also be conducted through condensation reactions, which usually serve for polyamide and polyesters synthesis. These reaction occurs among -OH or $NH_2$ groups and -COOH or derivatives respectively. N,N-(3-dimethylaminopropyl)-N-ethyl carbodiimide (EDC) is a highly efficient reagent for cross-linking hydrophilic polymers having amide groups.

3. **Crosslinking by Grafting:** It is the polymerization of a monomer on the backbone of a preformed polymer. The polymers chains are activated by pre-treatment like the action of chemical agents, or high energy radiation treatment. In chemical grafting, chemical reagents are used to activate macromolecular backbones. Hydrogels obtained showed an excellent pH-dependent swelling behavior and possess ideal characteristic to be used as drug vitamin delivery device in the small intestine [24]. High energy radiation such as gamma and electron beam is also used to initiate grafting. For example, Electron beam was used to initiate the free radical polymerization of acrylic acid on the backbone of CMC. Water radiolysis product will also be helpful to abstract proton form macromolecular backbones. Irradiation of CMC and monomer will produce free radicals that can combine to produce hydrogel.

### 7.5.3   CROSSLINKING BY RADIATIONS

The radiation crosslinking techniques are utilized for preparation of hydrogels of unsaturated compounds. The technique is based on producing free radicals in the polymer following the exposure to the high energy radiations such as gamma-ray, x-ray, or electron beam. Recombination of the macro-radicals on different chains lead to the formation of covalent bonds, so finally, a cross-linked structure is formed. It is a widely used technique since it neither requires chemical reagent nor initiator and the modification and sterilization can be achieved in a single step. The action of radiation (direct or indirect) will depend on the polymer environment (i.e., dilute solution, concentrated solution, solid state) [24].

Other technologies used in bio-responsive hydrogels preparation:

1. **Free Radical Polymerization:** This free radical polymerization method involves different steps like propagation, chain transfer, initiation, and termination steps. The initiation step requires radical's generation which reacts with the monomer converting them to active forms, that can be obtained by various initiators such as, thermal, UV, visible, and redox initiators. Acrylates, vinyl lactams, and amides are the most used monomers in this method which have proper functional groups that radically polymerize [25].

2. **Solution Polymerization:** Ionic or neutral monomers are mixed with the multiple functional crosslinking agents. The polymerization is started thermally by UV-irradiation or by redox initiator system. The advantage of solution polymerization is the solvent serves as a heat sink. The final product hydrogels is washed with distilled water to remove initiator, soluble monomers and oligomer, cross-linking agent and other impurities. Solvents used in this process are water-ethanol mixtures, water, ethanol, and benzyl alcohol [26].

3. **Suspension Polymerization or Inverse-Suspension Polymerization:** The product is obtained as powder or microspheres (beads) in this process. Water-in-oil (W/O) process is chosen in the polymerization so referred as inverse suspension. The monomers and initiators are distributed in the hydrocarbon phase homogenously. To govern the viscosity of the monomer solution, rotor design, agitation speed, and dispersant type, resin particle size and shape are used. Continuous agitation and addition of a low hydrophilic-lipophilic-balance (HLB) suspending agent are required as dispersion is thermodynamically unstable [26].

## 7.6 CHARACTERIZATION OF BIO-RESPONSIVE HYDROGELS

Characterization is the set of evaluations made for the final optimized product and to evaluate its parameters and finally to attest if it is good to use. Generally, hydrogels are characterized for their morphology, swelling property, chemical structure, and elasticity. The hydrogel characterization parameters are mentioned below.

### 7.6.1   pH

pH is measured using digital pH meter. The pH meter must be calibrated before used.

### 7.6.2   SOLUBILITY

According to Katayama et al. [27], the hydrogel content of a given material is estimated by measuring its insoluble part in a dried sample after immersion in deionized water for 16 h. The sample is prepared at a dilute concentration (around 1%), to ensure that hydrogel material is fully dispersed in water. The gel fraction is measured using the given below formula:

$$\text{Gel Fraction (hydrogel \%)} = (W_d/W_i) \times 100$$

where, $W_i$ is the initial weight of dried sample; $W_d$ is the weight of the dried insoluble part of sample after extraction with water.

### 7.6.3   SWELLING MEASUREMENT/SWELLING RATIO

To measure the swelling property of hydrogels the Japanese industrial standard K8150 method is usually used. According to this method, hydrogel is dried and then immersed in deionized water for 48h at room temperature on a roller mixer. The swelled hydrogel is then filtered through a standardized stainless steel net of 30 meshes (681 μm). The following formula is used to calculate the swelling.

$$\%\text{Swelling} = \{(W_s - W_d)/W_d)\}$$

where, $W_s$ is weight of swelled hydrogel; $W_d$ is the weight of dried hydrogel.
   Another alternative for swelling ratio measurement is carried out in a volumetric vial (Universal). The dry hydrogel (0.05 to 0.1g) was dispersed into sufficiently high quantity of water (25–30 mL) for 48 hours at ambient

temperature. The mixture is then centrifuged, the free water is removed, and the swelling ratio is calculated using the previous formula.

There is another standardized Japanese method called JIS K7223. In this method the dried gel is also immersed in deionized water but only for 16 hrs at room temperature. The process is similar to the first method except that the filtration is carried on in a stainless steel net of 100-meshes (149 μm). Swelling is calculated as follows:

$$Swelling = (C/B) \times 100$$

where, C is the weight of dried hydrogel and B is the weight of the insoluble portion after extraction with water [28].

### 7.6.4   *VISCOSITY/RHEOLOGY*

Rheological parameters of sample measure structural viscosity and kinetics of volatile components loss. Viscosity test is mostly performed using a cone plate digital rheometer (Brookfield viscometer) under constant temperature at 4°C [26].

The viscosity is calculated by the simple equation of angle of repose, through that height and length is determined.

### 7.6.5   *SCANNING ELECTRON MICROSCOPY (SEM)*

The scanning electron microscope (SEM) uses a focused beam of high-energy electrons to generate a variety of signals at the surface of solid specimens. SEM is used to reveal information about sample, surface morphology, chemical composition, and crystalline structure. Areas of approximately 1 cm to 5 microns in width can be imaged in a scanning mode using conventional SEM techniques (magnification ranging from 20X to approximately 30,000X, spatial resolution of 50 to 100 nm). SEM study is usually used also to determine pore size in the sample.

### 7.6.6   *FOURIER TRANSFORM INFRARED (FTIR) SPECTROSCOPY*

This technique is use to identify chemical structure of a substance by providing infrared spectrum of absorption or emission of a sample which can be a solid, a liquid or a gas. The principle behind is that, chemical bonds contained in the sample can absorb infrared light at specific frequencies. The obtained IR spectrum is a fingerprint of the measured sample. It is a highly solicited technique to investigate structural arrangements in hydrogels by comparison with the starting material.

### 7.6.7   LIGHT SCATTERING

Gel Chromatography or Gel permeation chromatography (GPC) is considered to be the most important test method for the determination of molecular properties of synthetic and natural macromolecules in solution. Distribution properties of macromolecules such as molecular weight distribution, structure distribution, branching distribution are easily obtained by this mean. GPC coupled on line to a multi angle laser light scattering (GPC-MALLLS) is a widely used technique to determine the molecular distribution and parameters of a polymeric system. For example, this technique is widely used in quantifying the hydrogels of several hydrocolloids such as gum Arabic, gelatine, and pullulan [29].

### 7.6.8   X-RAY DIFFRACTION

X-ray diffraction is carried out to determine morphological characteristics of the hydrogel. It gives the estimation of crystalline and amorphous characteristics. X-ray diffraction is used to understand whether polymers retain their crystalline nature or they get deform during pressurization process.

### 7.6.9   SPREADABILITY STUDY

It is a study carried out to evaluate the capacity of topical hydrogel to spread on the skin. The apparatus consists of wooden block with scale and two glass slides having a pan mounted on a pulley. The formulation is placed between two glass slides and 100 gms weight is placed on upper glass slide for 5 minutes. Weight can be added and the time required to separate the two slides is taken as spreadability time [27].

$$S = (m \times l) / t$$

where, S is spreadability; m is weight tied on upper slide; l is length of glass slide; and t is time taken in seconds.

## 7.7 MODERN APPLICATIONS OF BIO-RESPONSIVE HYDROGELS

According to Gough et al. [30], hydrogel materials are increasingly studied for applications in drug delivery, biological sensing, and tissue regeneration due the following reasons:

- Hydrogels provide suitable semi-wet, 3-D environment for molecular-level biological interactions;
- Many hydrogels provide antifouling property (prevent nonspecific adsorption of proteins);
- Biological molecules can be covalently incorporated into hydrogel structures;
- Mechanical properties of hydrogel are highly tunable.

### *7.7.1 APPLICATION OF BIO-RESPONSIVE HYDROGEL IN TISSUE ENGINEERING*

Tissue or organ transplantation is generally accepted therapy to treat patients suffering from tissue or organ loss or failure. However, this therapy showed some limitations such as trouble to find donor, sometimes incompatibilities problem and even rejection. An innovative strategy to treat patients in the need of new organs is tissue engineering.

Tissue engineering is a technology of regeneration of damaged cells or tissue in the body by using a combination of cells, engineering materials, and suitable chemical factors.

According to Langer and Vacanti [31], tissue engineering is defined as an interdisciplinary field of research that applies both the principles of engineering and the processes and phenomena of the life sciences toward the development of biological substitutes that restore, maintain, or improve tissue function [31].

The main purpose of tissue engineering is to regenerate living, healthy, and functional tissues that can be employed as tissues graft or organ replacement. The general approach is to utilize 3D scaffolds that function as temporary supports for cell growth and new tissue development.

Hydrogels with their crosslink network and other mechanical properties are ideal candidates to make scaffold for tissue engineering. There are various approaches to successfully engineer tissues or organs; however, the most employed strategy involves is based on the combination of patient's own cells with the polymer scaffold. These strategies involve the isolation of cells from patient's tissue by a small biopsy of the desired tissue's cells followed by the harvesting *in vitro*. Then the cells are incorporated in the 3D polymer scaffold where it acts as the natural ECM. The combined cells with the scaffold are introduced in the body, and the scaffold will deliver the cells to the desired site, it will provide a space for a new tissue formation, and will probably control the structure and function of the engineered tissue [32].

Many tissues and organs are being engineered using this approach, skin, bladder, cartilage, bone, ligament, etc. (Figure 7.5). For example, Oberpenning et al. [33] have reproducibly created *in vitro* transplantable urinary bladder neo-organs. Using urothelial and smooth tissue cells, grown in culture from canine native bladder biopsies and seeded onto preformed bladder shaped polymers; they have engineered a bladder that demonstrated normal capacity to retain urine, with normal elastic properties and histologic architecture during a functional evaluation up to 11 months [33].

More recently, Moutos et al. [34] have developed a method to grow a new cartilage of an arthritic joint by using a patient's stem cells. The technique is based on a 3-D biodegradable synthetic scaffold molded in the precise shape of the patient's joint. First of all, the cartilage is fabricated from patient's stem cells taken from fat beneath the skin, and then used to cover the scaffold. The whole system then is implanted onto the surface of an arthritic hip joint [34].

### 7.7.1.1   ATTRIBUTES FOR HYDROGELS IN TISSUE ENGINEERING

Numerous designs criteria are required for hydrogels to be suitably functional in tissue engineering or organs formation. These criteria include physical parameters and biological performance parameters. Some of the most important criteria are mentioned below.

Biocompatibility is defined by IUPAC as the ability to be in contact with a living system without causing any adverse effect. It is also defined as the ability of the material to exist in the body without destructing the adjacent cells.

The crosslinking is another important parameter need to control for a good result in bio-responsive hydrogel preparation. Three different mechanisms of gelling are usually observed; ionic crosslinking, covalent crosslinking, and inherent phase transition behavior. Ionic crosslinking presents the possibility of occurrence of uncontrolled deterioration of hydrogels properties, due to ions exchange with other ionic molecules in aqueous environment, while covalent crosslinking shows the risk of toxicity of crosslinking agent, and formation of non-degradable hydrogels. The utilization of phase transition behavior of some polymers is a recent approach to get good results.

The mechanical property of hydrogels is an important design parameter in tissue engineering. The gels must allow and facilitate tissue development but also adhesion and gene expression of the cells. Mechanical properties of hydrogels are driving other parameters also such as the type of crosslinking and the crosslinking density, rigidity of polymer chain, hydrophilic/hydrophobic balance.

Other critical parameters are controlled degradation of hydrogels which is depending to the type of tissue that has to be generated. Hydrogel degradation cannot be the same because that requires to control some factors directly related to the degradation rate and cells interaction with hydrogel affects their adhesion, migration, and differentiation.

### 7.7.1.2 ADVANTAGES AND DISADVANTAGES OF HYDROGELS IN TISSUE ENGINEERING

1. **Advantages:**

- Aqueous environment is adequate for protection of cells and fragile drugs such as protein, DNA, oligonucleotides, peptides.
- Usually biocompatible.
- Can be developed as *in situ* forming gel, i.e., a liquid, which will form gel at body temperature.
- Easy modification with cell adhesion.
- Good transport of nutrients to cells and products from cells.

2. **Disadvantages:**

- Sometime it can be hard to handle.
- Usually mechanical weak or too strong.

- Difficulty in loading of drugs and cells before proceeding to cross-link *in vitro* as prefabricated matrices.
- Difficulty in sterilization.

### 7.7.2 APPLICATION OF BIO-RESPONSIVE HYDROGEL AS MEDIATED DRUG DELIVERY

Drug delivery field has always been focused on controlled and targeted release of therapeutic agent with the purposes to enhance the efficiency and efficacy of therapeutic agent at specific sites of action with reduced undesirable effects on others cells. Nowadays 'smart' materials have gained a lot of interest and are considered as very efficient way to fulfill those purposes of drug delivery. These materials consist of novel biomaterials which respond to their environment and deliver the therapeutic agent on the basis of environmental signals of remote stimulus.

The limitations of conventional controlled release systems like incapacity to be sensitive to metabolic modification, inability to neither modulate drug release nor target specific cells have motivated the interest for bio-responsive polymers as drug carriers [35]. Bio-responsive polymers are able to respond to various types of stimuli. The bio-responsive polymers will react to subtle changes in the environment such as pH modification, the change of temperature in the target tissue, the change of concentration of some protein or peptides which is considered as the trigger or bio-stimuli that induce the release of the drug at the targeted site.

Some examples of bio-responsive hydrogels for controlled drug delivery are discussed in subsections.

### 7.7.2.1 pH-RESPONSIVE HYDROGELS

These hydrogels are sensitive to pH variation in the surrounding medium. This capacity is conferred to them by pH-responsive polymers which have been used for their preparation. pH-responsive polymers are able to react to solution pH by undergoing changes in their structure and property such as chain conformation, solubility. pH-responsive polymers can be defined as polyelectrolytes that has weak acidic or basic groups either accept (protonation) or release protons (deprotonation) in response to a change in the environmental pH [36]. These acidic and basic groups that characterize polymer responsive hydrogels are generally carboxyl, pyridine, sulfonic, phosphate, and tertiary amines.

The protonation and deprotonation mechanisms occurring on the functional group depend on pH solution and lead to various type of reactions from polymers. It may cause flocculation, chain collapse-extension, and precipitation, self-assembly such as gel formation, micelles formation, swelling, deswelling or surface-active behavior depending on the nature and architectural type of hydrogels. pH-responsive polymers can be from natural (dextran, HA, alginic acid, CS, and gelatin are among the most widely used) or synthetic origin, however, they consist of either acidic, basic or neutral groups. Weak Acidic/basic groups based pH-responsive polymers according to their pKa values accept protons at low pH and release them at high pH, that lead to a change in the ionic state (ionized/nonionized) which states influenced the solubility in aqueous solution. This transition results in swelling/deswelling of hydrogels.

Among pH-responsive acidic polymers, commonly preferred are carboxylic, sulfonic, phosphoric, and boronic acids (BA), sulfonic, and phosphoric acids in hydrogels preparation. The obtained hydrogels swell well under basic conditions, when pH solution is superior to their pKa. The most widely used polymers containing sulfonic acid are poly(2-acrylamido-2-methylpropane sulfonic acid) (PAMPS) and poly(4-styrenesulfonic acid) (PSSA) [36].

Concerning basic polymers utilized as pH-responsive polymers, they undergo ionic state transition, when solution pH is in between 7 to 11. The functional groups that are mainly used are groups that contain tertiary amines such as methacrylate, methacrylamide, and vinyl polymers. Tertiary amine methacrylate-based polymers such as PDMA, poly[(2-diethylamino) ethyl methacrylate] (PDEA), and poly[(2-diisopropylamino) ethyl methacrylate] (PDPA) are the most preferred species among the basic polymers. Particularly, PDMA is widely used weak basic polymer having not only pH-responsive nature but also thermo-responsive nature [30].

Other functional groups that can be exploited are imidazole, piperazine, pyrrolidine, and morpholino. For example, Poly[(2-N-morpholino) ethyl methacrylate] (PMEMA) is a morpholino group-containing polymer, widely used and that reacts to various stimuli such as pH, temperature, and ionic strength of the medium [36] (Figure 7.4).

➤ Some acidic pH responsive polymers:

Poly(phosphoric) acids            Poly(sulfonic) acids

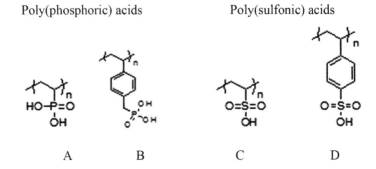

    A       B         C        D

➤ Some basic pH responsive polymers:

Tertiary amine groups      Morpholino group

    E       F        G

**FIGURE 7.4** Chemical structures of some pH responsive polymers; (A) PVPA poly (vinylphosphoric acid); (B) PVBPA poly(4-vinyl-benzyl phosphoric acid); (C) PVSA poly(vinylsulfonic acid); (D) PSSA poly(4-styrenesulfonic acid); (E) PDMA poly {(2-dimethylamino)ethyl methacrylate}; (F) PDPA poly[(2-diisopropylamino)ethylmethacrylate]; (G) PMEMA poly[(2-N-morpholino)ethyl methacrylate].

Xu et al. in 2015 [37] have developed a pH-sensitive hydrogel based on a natural polysaccharide (Hydroxypropyl Pachyman) crosslinked by epichlorohydrin (ECH-HPP). The hydrogel has been prepared and characterized to serve as an effective carrier for controlled delivery of proteins drugs bovine serum albumin (BSA) and lysozyme. ECH-HPP showed a high rate of entrapment percentage (97.6%) and a very low swelling ratio at gastric environment pH 1.2; while it showed a moderate swelling ratio at intestinal environment pH 7.4. The reported results indicate the great potential of the pachyman-based hydrogel as a safe, pH-responsive, and controllable protein drug delivery vehicle [37]. Xu et al. in 2018 [38] have developed a dual-drug loaded (doxorubicin (DOX) and tetracycline) pH-responsive hydrogel, based on poly(L-lactide)-co-polyethyleneglycol-co-poly(L-lactide) dimethacrylate as a macromolecular crosslinker and copolymerized with acrylic acid and N-isopropylacrylamide (MA-PLLA-PEG-PLLA-MA, AA, and NIPAM) for pH-controlled drug release. The hydrogels were pH-sensitive and shrinked at pH 1.2, while swelled and released most of their content at pH 7.4 [38].

## 7.7.2.2 THERMO-RESPONSIVE HYDROGELS

The change from ambient temperature to physiological temperature as well as the variation of body temperature observed in many disease or pathological conditions can be exploited in biomedical applications. Temperature variations have been exploited as a physical stimulus which acts as a trigger for delivery of therapeutics in thermo-responsive drug delivery systems. Thermo-responsive hydrogels modulates their gelation behavior according to temperature changes. They exhibit swelling behavior according to the change in response to surrounding temperature. These systems showed the advantage of forming gel *in situ*. As soon as biomaterial solution move from outside to inside the body, there is an instantaneously gel formation once temperature is reached, without need of chemical initiator (the solution is delivered in a minimally invasive manner and will solidify inside the body) [39]. Thermo-responsive hydrogels can be either physical gel or covalently linked gel that is loaded with drug at room temperature. Covalently linked networks will show a change in the degree of swelling in response to temperature variation, while physical gel will exhibit a sol-gel transition.

Thermo-responsive polymers are classified into two categories: (a) Thermo-responsive polymers with lower critical solution temperature (LCST); (b) thermo-responsive polymers with upper critical solution temperature (UCST). In fact, temperature-sensitive polymers have a specific critical solution temperature at which miscibility change observed between polymer and solvent. According to it, critical solution temperature can be set by UCST or LCST.

LCST and UCST are the respective critical temperature points below and above which the polymer and solvent are completely miscible [40].

In different words, UCST polymers become soluble after heating and LCST polymers become insoluble after heating. For example, below the UCST, the polymer presents a cloudy aspect, and the mixture polymer with solvent is non-miscible (two phase) (Figure 7.5).

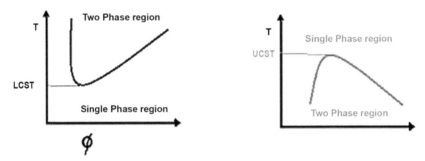

**FIGURE 7.5**   Polymer solution LCST and UCST behaviors according to temperature and volume fraction.

The most intensively studied of them is poly(N-isopropyl acrylamide) (PNIPAAM), which has his LCST close to body temperature around 32°C. Hydrogels obtained from these polymers (LCST), get hydrated and swell under the critical temperature, while collapse above it.

Many studies about development of thermo-responsive hydrogels based PNIPAAM for controlled drug release has been reported.

Tan et al. in 2009 [41] have developed a series of thermo-responsive hydrogels made of poly(N-isopropylacrylamide) coupled with aminated HA through amide bond linkage (AHA-g-PNIPAAm) for adipose tissue engineering [41].

Lu et al. have prepared biodegradable, thermo-responsive hydrogel for the sustained release of levofloxacin. The hydrogel consisted of (N-isopropylacrylamide) NIPAM and two biodegradable crosslinkers poly(ε-caprolactone) dimethacrylate (PCLDMA) and bisacryloylcystamine (BACy). The hydrogel showed slowly degradation in glutathione (GSH) at physiological pH and 37°C of temperature [42].

Liu et al. in 2017 [43] have developed an injectable thermo-responsive hydrogel made of Alginate-g-poly(N-isopropylacrylamide) as a smart drug delivery system for controlled release of DOX encapsulated micelles. (NIPAM) with different molecular weight was synthesized by atomic transfer radical polymerization and has been combined with alginate. They reported that at higher concentration (7.4 wt%) the copolymer formed solution at 25°C, when temperature increased to the body temperature (37°C) turn to hydrogel [43].

### 7.7.2.3 ROS RESPONSIVE HYDROGELS

Reactive oxygen species (ROS) are oxidant molecules derived from oxygen ($O_2$) which are either free radicals (superoxide radicals, nitric oxide radicals, hydroxyl radical (OH), alkoxy radicals (RO), peroxyl radicals) or molecular species capable of generating free radicals (hydrogen peroxide ($H_2O_2$), peroxynitrite (ONOO-), hypochlorous acid (HOCl)). They widely exist in living organisms and play important role in cell signaling pathway. However, an overproduction of oxidants molecules leads to disruption of cellular homeostasis and give rise to oxidative stress and related diseases, especially inflammatory diseases. For example, $H_2O_2$ level in normal human plasma is $1-8\times10^{-6}$ M (with an average of $\approx 3\times10^{-6}$ M), while activated macrophages may generate a higher local $H_2O_2$ concentration in the range of $10-1000 \times10^{-6}$ M. The unique redox microenvironments make the pathological regions distinct from their surroundings, and can be exploited as symbols for disease detection, diagnosis, and therapy [44].

This ROS responsive technology is based on the use of polymeric materials that show sensitivity to reactive oxygen species. Most of oxidation sensitive materials studied consist of entities such as thioether, selenium, tellurium, thioketal, aryl BA. Their interaction with oxidants molecules induced change in polymer properties such as degradation of polymeric materials or solubility switch.

### 7.7.2.3.1 Selenium and Tellurium

They are organochalcogen family's members. Selenium is an essential trace element present in animals and humans found in a vital enzyme in the body which is GSH peroxidase that plays a very important role against oxidative stress as an antioxidant enzyme. Selenide or diselenide containing polymers are more sensitive to mild conditions compared to sulfur-containing counterparts that is gaining researcher's attention to develop ROS responsive materials (Figure 7.5). The hydrophobic monoselenide can be converted in selenoxide or selenone under oxidative conditions [45].

Zhang et al. have developed a dual redox responsive block copolymer containing diselenide bonds, which reacted to oxidant and reductant in solution even at a very low concentration [46].

Cao and co-workers have reported a novel tellurium containing polymer micelle system; ultrasensitive to $H_2O_2$ (100µM) and which could undergo series of morphological changes [42] (Figure 7.6).

**FIGURE 7.6**   Schematic illustration of selenium and tellurium oxidation.

### 7.7.2.3.2   *Aryl Boronic Ester/Aryl Boronic Acid (BA)*

Aryl BA and ester bonds have been extensively used in designing functional materials, especially as sensors because to their binding affinities for diols, amino alcohols, cyanide, and fluoride. Aryl boronic ester bonds are oxidized in phenols and BA in the presence of $H_2O_2$ molecule. That makes Arylboronic ester-containing polymers, interesting candidate for ROS responsive drug delivery. Despite their sensitivity to $H_2O_2$, they cannot be degraded by other ROS. They are important in the construction or synthesis of reactive polymers and organic compounds.

Frechet and coworkers have synthesized an oxidation sensitive polymer-based dextran which is a water-soluble, biocompatible polymer of glucose. They replaced hydroxyls groups at one end of the polymer with Aryl boronic ester bonds, such that the polymer became water-insoluble, but solubilizable in appropriate organic solvent. That modified dextran used for the preparation of microparticles oxi-dex of 100 nm, which degraded at a half-life of 36 min in a solution concentrated at 1 mM $H_2O_2$ compared to a half-life of more than 1 week in absence of $H_2O_2$ (Figure 7.7) [48].

ROS responsive hydrogels can be used as local drug depots that attempt to achieve localized and controlled drug delivery. Under oxidative stress, drug-loaded ROS sensitive hydrogels may release in an accelerated manner.

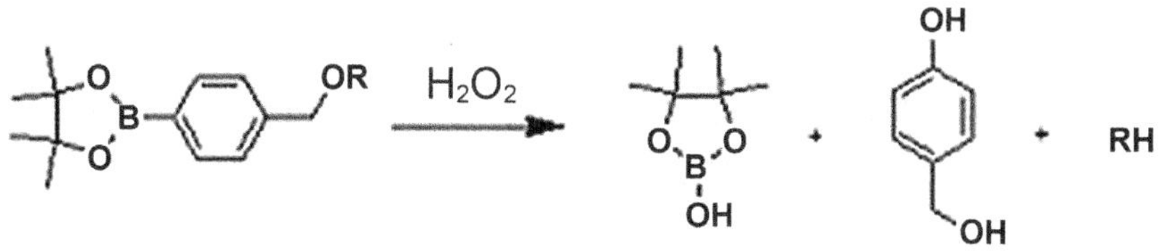

**FIGURE 7.7**  Illustration of aryl boronic acid degradation in presence of $H_2O_2$.

Duvall and co-workers have synthesized an ABC triblock polymer poly[(propylene sulfide)-block-(N,N-dimethyl acrylamide)-block-(N-isopropyl acrylamide)](PPS-b-PDMA-b-PNIPAAM) by a combination of anionic and RAFT polymerization, which has been used to developed a thermo-responsive hydrogel with ROS-triggered degradation and drug release. The polymer formed a physical crosslinked hydrogel when transitioned from ambient to physiologic temperature. When exposed to ROS, the PPS block polymer became hydrophilic and end up by the degradation of the hydrogel. The hydrogel also showed a cell protection against ROS and a $H_2O_2$ dose release profile [49]. ROS-responsive polymeric scaffolds and hydrogels are capable of sensing the oxidative stress in the surrounding environment, modulating cell behavior, facilitating on-demand drug release, and eliminating excess ROS.

## 7.7.3  BIORESPONSIVE HYDROGEL FOR BIOSENSING APPLICATION

These types of responsive materials are designed to provide higher-order responsiveness to biological stimuli which are directly sensitive to specific molecules such as antigen-antibody binding or enzyme-substrate interactions. In these systems, analytes recognition events are coupled to a macroscopically observable change in the polymer, such as simple expansion or contraction of polymer network, fluorescence response change of fluorophore in the gel, a change in the diffracted wavelength in a colloidal assembly, a change in the optical properties of the gel [50].

### 7.7.3.1  PHYSICAL EXPANSION OR CONTRACTION RESPONSE

Physical expansion is the basis of change for all hydrogel systems. It is characterized by a physical change observable in the size of the gel due to a change in network density. Many changes such as size, porosity, density, refractive index, and modulus may occur on the hydrogel following its swelling that rends this sensing modality particularly useful. A wide range of sensor transduction method is applied to hydrogel-based bioresponsive materials. But for many bioresponsive hydrogels, this remains a relatively untapped area of research, with far more effort having been spent on the development of new materials than on their real-world application in sensors [50].

Miyata et al. [51] have prepared an antigen-sensitive hydrogel by using antigen-antibody binding at a crosslinking point in the hydrogel. The hydrogel was able to exhibit swelling changes in response to antigen recognition. They focused on antigen-antibody complex formation because specific recognition of an antibody can provide the basis for fabricating sensing devices with a wide range of applications in immunoassay and antigen sensing [51].

### 7.7.3.2   FLUORESCENCE RESPONSE

In this modality, hydrogels are integrated with fluorescents tags to provide an easily detectable result of any change in the hydrogel network.

Russell et al. in 1999 [52] have prepared a fluorescence-based glucose biosensor using concanavalin A and dextran as an entrapped drug in poly(ethylene glycol) hydrogel. A photopolymerized PEG hydrogel containing fluorescein isothiocyanate dextran (FITC-dextran) and tetramethylrhodamine isothiocyanateconcavalin A (TRITC-Con A) were chemically conjugated into the hydrogel using an alpha-acryloyl, omega-N-hydroxysuccinimidyl ester of PEG-propionic acid. In the presence of glucose which binds to TRITC-Con A, FITC-dextran is liberated, resulting in increased FITC fluorescence proportional to the glucose concentration [52].

### 7.7.3.3   DIFFRACTION RESPONSE

In this system, hydrogels have been coupled to the diffraction from photonic crystals. It involves encapsulation of a photonic crystal inside a network hydrogel or by the direct assembly of photonic crystals using microgels as buildings blocks. The advantage of sensing the construct system is that Bragg diffraction they show is responsive to hydrogel swelling.

Ben-Moshe et al. [53] have developed a new photonic crystal polymerized crystalline colloidal array (PCCA) glucose sensing materials. These materials are composed of hydrogels that incorporate an array of ~100-nm-diameter monodisperse polystyrene colloids that Bragg diffracts light in the visible spectral region. The hydrogels change the volume with the variation of the glucose concentration. This changes the lattice spacing, which changes the wavelength of the diffracted light [53].

## 7.8  FUTURE OUTLOOKS

In conclusion, hydrogel-based networks have been fabricated and modified to achieve the desired properties and useful in different biomedical applications. The need for easy administration in organ/tissue regeneration medical applications drove the cell response suitable for minimally invasive treatments in the research of injectable hydrogels and bio-responsive constructs. When put in contact with aqueous solution hydrogels has the ability to swell. The present detail indicates about the classification of hydrogels on different bases, physical, and chemical characteristics and technical feasibility of their utilization, method of preparation and biomedical application. Bio-responsive hydrogels will have a good scope in future applications as the next-generation biomaterials. Thus, hydrogels are called a smart or intelligent biomaterial.

## ACKNOWLEDGMENT

I am grateful to Dr. Pinkal Patel and Mr. Nadim Chippa for their help to draw the chemical structure in the book. I would like to thank all the contributors to complete this book on time.

## KEYWORDS

- carboxymethyl cellulose
- differential scanning calorimetry
- extracellular matrix
- interpenetrating polymeric hydrogel
- matrix metalloproteinase
- polyvinyl alcohol

## REFERENCES

1. Lee, S. C., Kwon, I. K., & Park, K., (2013). Hydrogels for delivery of bioactive agents: A historical perspective. *Adv. Drug Deliv. Rev., 65*, 11–20.
2. Kopecek, J., (2007). Hydrogel biomaterials: A smart future? *Biomat., 28*, 5181–5192.
3. Buwalda, S. J., Boere, K. W., Dijkstra, P. J., Feijen, J., & Vermonden, T., (2014). Hydrogels in a historical perspective: From simple networks to smart materials. *J. Cont. Rel., 190*, 251–273.
4. Yahia, L. H., Chirani, N., & Gritsch, L., (2015). History and applications of hydrogels. *J. Biomedical Sci., 4*, 2.
5. Sgambato, A., Cipolla, L., & Russo, L., (2016). Bioresponsive hydrogels: Chemical strategies and perspectives in tissue ENGG. *Gels, 2*, 28.
6. Necas, J., et al., (2008). Hyaluronic acid (hyaluronan), a review. *Veterinarni Medicina, 53*(8), 391–411.
7. Ahmadi, F., Oveisi, Z., Samani, S. M., & Amoozgar, Z., (2015). Chitosan based hydrogels: Characteristics and pharmaceutical applications. *Res. Pharm. Sci., 10*, 1–16.
8. Augst, A. D., Kong, H. J., & Mooney, D. J., (2006). Alginate hydrogels as biomaterials. *Macromolecular Bioscience, 6*(8), 623–633.
9. Jia, X., & Kiick, K. L., (2009). Hybrid multicomponent hydrogels for tissue engineering. *Macromol. Biosci., 9*(2), 141–156.
10. Lee, H. J., (2015). "Hybrid hydrogels for tissue engineering." *All Dissertations*, 1775.
11. Nam, K., Kimura, T., & Kishida, A., (2007). Preparation and characterization of cross-linked collagen phospholipid polymer hybrid gels. *Biomaterials, 28*, 1.
12. Grieshaber, S. E., Farran, A. J., Lin-Gibson, S., Kiick, K. L., & Jia, X., (2009). Synthesis and characterization of elastin-mimetic hybrid polymers with multiblock, alternating molecular architecture and elastomeric properties. *Macromol., 42*(7), 2531–2541.
13. Takashi, L., Hatsumi, T., Makoto, M., Takashi, I., Takehiko, G., & Shuji, S., (2007). Synthesis of porous poly(N-isopropylacrylamide) gel beads by sedimentation polymerization and their morphology. *J. Appl. Polym. Sci., 104*(2), 842.
14. Yang, L., Chu, J. S., & Fix, J. A., (2002). Colon-specific drug delivery: New approaches and *in vitro/in vivo* evaluation. *Int. J. Pharm., 235*, 1–15.
15. Maolin, Z., Jun, L., Min, Y., & Hongfei, H., (2000). The swelling behavior of radiation prepared semi-interpenetrating polymer networks composed of polyNIPAAm and hydrophilic polymers. *Rad. Phys. Chem., 58*, 397–400.
16. Hacker, M. C., & Mikos, A. G., (2011). *Synthetic Polymers, Principles of Regenerative Medicine* (2nd edn., pp. 587–622).
17. Garg, S., & Garg, A., (2016). Hydrogel: Classification, properties, preparation, and technical features. *Asian J. Biomat. Res., 2*, 161–170.
18. Griffith, L. G., (2000). Polymeric biomaterials. *Acta Materialia, 48*, 261–277.
19. Dong, L. C., Hoffman, A. S., & Yan, Q., (1994). Dextran permeation through poly(N-isopropylacrylamide) hydrogels. *J. Biomat. Sci., 5*, 471–484.
20. Nilimanka, D., (2013). Preparation methods and properties of hydrogel: A review. *Int. J. Pharma. Sci., 5*(4).
21. Mondal, M., Trivedy, K., & Nirmal, K. S., (2007). The silk proteins, serin and fibroin in silkworm bombyxmori. *Caspian J. Env. Sci., 5*, 61–76.

22. Gulrez, S. K. H., Al-Assaf, S., & Phillips, G. O., (2011). Hydrogels: Methods of preparation, characterization, and applications. *Prog. Mol. Env. Bioengg.*

23. Nolan, A., Anthony, G. E., (2013). Bioresponsive hydrogels. *Adv. Healthc. Mater, 2*(4), 521–32.

24. Syed K. H., Gulrez, S., Al-Assaf, & Glyn, O. P., (2011). Hydrogels: Methods of preparation, characterization, and applications. *Prog. in Mol. Env. Bioengg. Angelo. Carpi., Intech. Open.*

25. Shuanhong, M., Bo, Y., Xiaowei, P., & Feng, Z., (2016). Structural hydrogels. *Polymer, 98,* 511–535.

26. Singh, S. K., Dhyani, A., & Juyal, D., (2017). Hydrogel: Preparation, characterization and applications. *Ph. Inn. J., 6*(6), 21–32.

27. Katayama, T., Nakauma, M., Todoriki, S., Phillips, G. O., & Tada, M., (2006). Radiation-induced polymerization of gum Arabic (Acacia sengal) in aqueous solution. *Food Hydrocolloids, 20,* 981–989.

28. Qavia, S., Pourmahdiana, S., & Eslamia, H., (2014). Acrylamide hydrogels preparation via free radical crosslinking copolymerization: Kinetic study and morphological investigation. *J. Macromol. Sci., Part A: Pure and Applied Chemistry, 51,* 841–848.

29. Mansur, H. S., Orefice, R. L., & Mansur, A. A. P., (2004). Characterization of poly(vinyl alcohol)/poly(ethylene glycol) hydrogels and PVA-derived hybrids by small-angle x-ray scattering and FTIR spectroscopy. *Polymer, 45,* 7191–7202.

30. Ulijn, R. V., Bibi, N., Jayawarna, V., Thornton, P. D., Todd, S. J., Mart, R. J., Smith, A. M., & Gough, J. E., (2007). Bioresponsive hydrogels. *Materials Today, 10*(4), 41–48.

31. Langer, R., & Vacanti, J. P., (1993). Tissue engineering. *Science, 260*(5110), 921–926.

32. Lee, K. Y., & Mooney, D. J., (2001). Hydrogels for tissue engineering. *Chem. Reviews, 101*(7), 1869–1880.

33. Oberpenning, F., Meng, J., Yoo, J. J., & Atala, A., (1999). De novo reconstitution of a functional mammalian urinary bladder by tissue engineering. *Nat. Biotech., 17,* 149–155.

34. Moutos, F. T., Glass, K. A., Compton, S. A., Ross, A. K., Gersbach, C. A., Guilak, F., & Estes, B. T., (2016). Anatomically shaped tissue-engineered cartilage with tunable and inducible anti-cytokine delivery for biological joint resurfacing, *Proc. Natl. Acad. Sci. U.S.A., 2, 113*(31), E4511–4522.

35. Jin, Y., Dariela, A., George, J. Y., & Debra, T. A., (2010). Bioresponsive matrices in drug delivery. *J. Bio. Engg., 4,* 15.

36. Kocak, G., Tuncer, C., & Butun, V., (2017). pH- Responsive polymers. *Pol. Chem., 8*(1), 144–176.

37. Xu, W., He, X., Zhong, M., Hu, X., & Xiao, Y., (2015). A novel pH-responsive hydrogel based on natural polysaccharides for controlled release of protein drugs. *RSC Advances, 5*(5), 3157–3167.

38. Xu, L., Qiu, L., Sheng, Y., Sun, Y., Deng, L., Li, X., & Zhang, R., (2018). Biodegradable pH-responsive hydrogels for controlled dual-drug release. *Journal of Materials Chemistry B., 6*(3), 510–517.

39. Klouda, L., (2015). Thermoresponsive hydrogels in biomedical applications. *Eur. J. Pharm. Biopharm., 97,* 338–349.

40. Ward, M. A., & Georgiou, T. K., (2011). Thermoresponsive polymers for biomedical applications. *Polymers, 3*(3), 1215–1242.

41. Tan, H., Ramirez, C. M., Miljkovic, N., Li, H., Rubin, J. P., & Marra, K. G., (2009). Thermosensitive injectable hyaluronic acid hydrogel for adipose tissue engineering. *Biomaterials, 30*(36), 6844–6853.

42. Gan, J., Guan, X., Zheng, J., Guo, H., Wu, K., Liang, L., & Lu, M., (2016). Biodegradable, thermoresponsive PNIPAM-based hydrogel scaffolds for the sustained release of levofloxacin. *RSC Adv., 6*(39), 32967–32978.

43. Liu, M., Song, X., Wen, Y., Zhu, J. L., & Li, J., (2017). Injectable thermoresponsive hydrogel formed by alginate-g-poly(N-isopropylacrylamide) that releases doxorubicin-encapsulated micelles as a smart drug delivery system. *ACS Applied Mat Interfaces, 9*(41), 35673–35682.

44. Xu, Q., He, C., Xiao, C., & Chen, X., (2016). Reactive oxygen species (ROS) responsive polymers for biomedical applications. *Macromol. Biosci., 16*(5), 635–646.

45. Saravanakumar, G., Kim, J., & Kim, W. J., (2016). Reactive-oxygen-species-responsive drug delivery systems: Promises and challenges. *Adv. Sci., 4*(1), 1600124.

46. Ma, N., Li, Y., Xu, H., Wang, Z., & Zhang, X., (2010). Dual redox responsive assemblies formed from diselenide block copolymers. *J. American Chem. Soc., 132*(2), 442, 443.

47. Cao, W., Gu, Y., Li, T., & Xu, H., (2015). Ultra-sensitive ROS-responsive tellurium-containing polymers. *Chem. Comm., 51*(32), 7069–7071.

48. Broaders, K. E., Grandhe, S., & Fréchet, J. M. J., (2011). A biocompatible oxidation-triggered carrier polymer with potential in therapeutics. *J. American Chem. Soc., 133*(4), 756–758.

49. Gupta, M. K., Martin, J. R., Werfel, T. A., Shen, T., Page, J. M., & Duvall, C. L., (2014). Cell Protective, ABC triblock polymer-based thermoresponsive hydrogels with ROS-triggered degradation and drug release. *J. American Chem. Soc., 136*(42), 14896–14902.

50. Hendrickson, G. R., & Andrew, L. L., (2009). Bioresponsive hydrogels for sensing applications. *Soft Matter, 5*(1), 29–35.

51. Miyata, T., Asami, N., & Uragami, T., (1999). Preparation of an antigen-sensitive hydrogel using antigen-antibody bindings. *Macromol., 32*(6), 2082–2084.

52. Russell, R. J., Pishko, M. V., Gefrides, C. C., McShane, M. J., & Coté, G. L., (1999). A fluorescence-based glucose biosensor using concanavalin A and dextran encapsulated in a poly(ethylene glycol) hydrogel. *Anal. Chem., 1, 71*(15), 3121–3132.

53. Ben-Moshe, M., Alexeev, V. L., & Asher, S. A., (2006). Fast responsive crystalline colloidal array photonic crystal glucose sensors. *Anal Chem., 78*(14), 5149–5157.

# Index

## 1

1,3-bis (2-chloroethyl)-1-nitrosourea (BCNU), 104
1,4-butanediol diglycidylether (BDDGE), 200

## β

β-cyclodextrin, 19, 96
β-galactosidase (β-gal), 191

## ω

ω-alkoxamine, 43
ω-dithioester, 43
ω-halide, 43

## A

acrylamide, 4, 28, 44, 45, 60, 100, 131, 133, 138, 227
acrylate, 9, 44, 47, 83
acrylic acid, 7, 29, 31, 57, 75, 82, 95, 96, 139, 201, 212, 223
active moieties, 4, 58, 161
adeno-associated virus (AAV), 122
adenocarcinoma tumors, 105
adenosine, 20, 141
  diphosphate, 141, 144
  triphosphate (ATP), 7, 20–22, 141, 158
  responsive polymers, 141
adenovirus (AV), 122
  complexes, 142, 143
adhesive
  properties, 74
  technology, 77
age-related macular degeneration (AMD), 93, 94, 108
albino rabbits, 95, 96, 97
alkoxy radicals (RO), 137, 225
alpha-chemotrypsin, 141

amino
  acids, 47, 50, 75, 131, 140
  groups, 29, 30, 125, 126, 135
amphiphiles, 8, 10, 49
amphiphilic
  block copolymers, 8, 48, 51, 52
  copolymers calorimetric techniques, 55
  molecules, 179
angiogenesis, 13, 22, 188
anionic
  hydrocolloids, 210
  polymerizations, 41
  polymers, 8, 80
  polysaccharide, 88
antibiotic, 84
  coumermycin, 84
  drugs, 175
  resistant microbes, 175
anticancer
  drugs, 23, 102, 103
  toxins, 154
antifungal activities, 97
antigen
  epitopes, 103
  responsive polymers, 5
anti-inflammatory
  cargo, 11
  drugs, 12
anti-Parkinsonism, 87
antitumor activity, 103, 144
apolipoproteins, 174, 180–182
arginine-grafted
  poly(cystaminebisacrylamide diamino-hexane) (ABP), 136
aromatase inhibitors, 152
arteriosclerosis, 10
arthrosclerosis, 137
asialoglycoprotein (ASGP), 153, 191
atom transfer radical polymerization (ATRP), 41–46, 63, 127

atomic force microscopy, 99
attenuated total reflection (ATR), 51, 63
azoaromatic bonds, 61
azobenzene (AZO), 22, 189
azoreductase, 60, 61

**B**

bioactive
  compound, 30
  molecules, 48, 62, 79, 162
  species, 159
bioactivity, 159, 202
biochemical
  processes, 83, 166
  sciences, 3
  stimulus, 83
biodegradability, 8, 20, 126, 131, 136, 184, 198, 200
biodegradation, 139, 181
bio-engineering, 42
biological
  agent, 57, 58, 158, 159, 208
  barriers, 180
  coating technologies, 2
  degradation, 21
  rhythms, 5
  signals, 1, 12
  stimuli/stimulus, 6, 58, 74, 227
biomacromolecules, 48, 158, 162, 199
biomaterials, 1, 108, 186, 197, 200, 220, 229
biomedical applications, 5, 14, 153, 163, 166, 174, 184, 201, 207, 223, 229
  devices, 1, 153
biomolecules, 2, 14, 25, 28, 156–158, 173, 178–180, 184, 185, 198, 208
  biomolecular drugs, 107
biopharmaceuticals, 79, 85, 90
biopolymers, 73, 166, 199
biorecognition, 158, 161
  moiety, 57
  species, 58, 158, 159
bio-responsive
  block copolymers, 41, 42
  copolymer gels, 51
  delivery, 77

drug, 3, 14, 79, 98, 131
  fluorescent, 158
  hydrogels, 25, 158–160, 163, 164, 208
    biosensing, 227
    biospecificity conferring, 161
    characterization, 213
    chemical crosslinking, 211
    controlled closed loop theory, 160
    Fourier transform infrared (FTIR) spectroscopy, 215
    light scattering, 216
    mediated drug delivery, 220
    modern applications, 217
    next generation hydrogels, 159
    pH, 214
    physical crosslinking, 209
    preparation, 209
    radiations crosslinking, 212
    scanning electron microscopy (SEM), 215
    solubility, 214
    spreadability study, 216
    swelling measurement/swelling ratio, 214
    tissue engineering, 217
    viscosity/rheology, 215
    x-ray diffraction, 216
  linkers, 27, 126
  materials, 1, 12, 25, 228
    engineering particulate moieties, 153
  nanoparticles, 173
    biological barriers, 179
    biological strategies, 179
    improving drug efficacy, 174
    opsonization and degradation, 181
  polymeric
    materials, 3
    systems, 5
  polymers, 3, 5, 24–26, 41, 58, 123, 124, 142, 220
    antigen responsive polymer, 62
    application, 24
    atom transfer radical polymerization (ATRP), 43
    biomedical applications, 155
    bioresponsive polymers, 156
    characterization, 50

cryogenic transmission electron
microscopy (CRYO-TEM), 54
DNA responsive polymer, 62
enzyme responsive polymer, 60
enzymes, 12
gene therapies, 156
glucose responsive polymer, 58
glucose, 15
hypoxia, 22
inflammation-responsive polymers,
61
ions, 18
living radical polymerization (LRP),
42
mechanical cues, 24
mechanical properties, 56
mechanism, 57
molecular imaging, 156
N-carboxy anhydride (NCA) polymerization, 47
nitroxide mediated radical polymerization (NMRP), 41, 42, 45, 46
optical probes, 157, 158
oral drug delivery, 79
parenteral drug delivery, 97
pH critical point, 57
pH sensitive polymers, 5
pH-responsive polymers, 62
polymer phase properties, 55
polymers, 156
redox responsive polymer, 62
redox, 10
reversible addition-fragmentation
chain transfer (RAFT), 42, 43, 45,
46, 127, 227
ring-opening metathesis polymerization (ROMP), 42, 47, 48
scattering techniques, 52
separation techniques, 51
spectroscopic techniques, 51
swelling measurements, 56
synthesis, 41
temperature, 23
topical drug delivery, 74
systems, 74, 75, 79, 81, 108, 140
vehicles, 155
biosafety, 187, 208

biosensing, 2, 73, 159, 173
biosensors, 11, 157, 161–163, 228
biotherapeutic agent, 124
bisacryloylcystamine (BACy), 224
blood
brain barrier, 87
coagulation, 13, 14
glucose levels (BGLs), 15, 18, 32, 107
thrombin, 13
boronic
acid (BA), 15, 16, 18, 30, 48, 49, 139,
141, 221, 225–227
ester groups, 10
bovine serum albumin (BSA), 60, 81, 94,
222
Box-Behnken experimental model, 94
brain
derived neurotrophic factor (BDNF),
91, 108
drug targeting, 85
Brownian motion, 52

**C**

camptothecin, 18, 104, 188
cancer
cell proliferation, 13
immunotherapeutic delivery system, 16
therapeutics
bio-responsive drug delivery, 151,
152
therapy, 141, 179, 186, 187
carbohydrates, 48, 211
carboxyl groups, 8, 9, 60, 62, 80, 210
carboxylic acid, 11, 19, 29, 61, 102
carboxymethyl
cellulose (CMC), 210, 212, 229
dextran (CM-Dex), 23, 32
cardiovascular
disease, 11, 157
pathologies, 79
cargo, 13, 17, 19, 87, 125, 173, 177, 181,
186, 187
catalase (CAT), 137, 189, 190, 193
cationic
copolymers, 16
lipid, 89

polymerizations, 41
polymers, 19, 27, 80
cell
  binding
    activity, 25
    sites, 25
  line studies, 102
  lymphoma models, 102
  metabolism, 21
  monolyers, 88
cellular
  antioxidants, 137
  endocytosis, 104
  internalization, 130, 158, 180, 190
  membrane, 127, 174, 181
  networks, 160
  receptors, 123
  toxicity, 26, 86
cellulose, 4, 99
chain transfer agent (CTA), 45, 63
charge conversion property (CCP), 130,
  144
chemotherapeutics, 178
chemotherapy, 100–104, 107, 152, 192
chimeric antigen receptors (CARs), 155,
  166
chitosan (CS), 4, 9, 13, 20, 28–30, 49,
  59–61, 76, 82, 87–90, 93, 100, 101, 108,
  125, 131, 133, 136, 137, 140, 183, 186,
  199, 200, 202, 210, 221
  modified PLGA (CS-PLGA), 90
chronic
  inflammatory diseases, 137
  oxidative stress, 137
chronotherapy, 5
chymotrypsin, 49, 158
cisplatin, 83, 104, 105, 107
collagen, 199, 200, 202, 203
colloids, 53, 229
complex coacervation, 207, 210
concanavalin A (Con A), 15, 17, 18, 31,
  32, 60, 228
controlled radical polymerizations (CRPs),
  41, 63
copolymer, 8–10, 14–16, 22, 24, 28, 41,
  42, 44–46, 49–52, 54–57, 76, 78, 80,
  82–84, 89, 99–102, 106, 107, 128, 129,

132–140, 143, 152, 153, 187, 200, 201,
  224, 226
copolymerization, 24, 85, 133, 209, 210
core-crosslinked redox responsive
  nanoparticles (CC-RRNs), 192
covalent bonds, 26, 185, 212
critical
  micellization temperature (CMT), 55, 56
  parameters, 28, 219
cryogenic transmission electron microscopy (Cryo-TEM), 50, 54, 55
crystallinity, 30, 53
cyclodextrin, 49, 60, 140
cystaminebisacrylamide (CBA), 136, 138,
  143
cysteine (Cys), 49, 134, 144
cytomegalovirus (CMV), 142
cytoplasm, 90, 122, 124, 191
cytosol, 10, 22, 26, 91, 102, 124, 126, 130,
  135, 174, 191
cytotoxicity, 15, 18, 23, 26, 27, 29, 106,
  126, 127, 130, 131, 133, 136, 141, 166,
  207, 208

**D**

deacetylation, 20, 88, 200
dendrimers, 9, 131, 136, 166, 189, 190
dendritic cells (DCs), 75, 154, 173, 176,
  177, 193
deoxyribonucleic acid (DNA), 49, 8, 21,
  22, 49, 58, 62, 63, 79, 101, 103, 122,
  123, 125, 129, 130, 133–139, 156, 187,
  219
  minicircles (mcDNA), 135, 136
deprotonation, 9, 29, 62, 83, 124, 220, 221
dextran, 10, 13, 17, 49, 60, 140, 183, 192,
  210, 221, 226, 228
diamminedichlorodisuccinato-platinum
  (DSP), 10, 138
diblock copolymers, 54, 55
differential scanning calorimetry (DSC),
  51, 55, 205, 229
dimethylamino ethyl methacrylate
  (DMAEMA), 9, 44, 59, 60, 76, 129,
  133, 134, 162

disulfide bond (-SS-), 10, 12, 49, 126, 134–136, 138, 142, 178, 190–192
dithiothreitol (DTT), 26, 32, 136, 192
docetaxel (DTX), 11, 92, 102
domino effect, 60
Donnan effects, 206
doxorubicin (DOX), 8, 13, 21–23, 32, 83, 100–104, 132, 135, 136, 152, 153, 187, 189, 190, 192, 193, 223, 224
drug
  carriers, 2, 49, 102, 104, 192, 220
  delivery
    carriers, 13
    systems, 2, 17, 18, 20, 27, 85, 186, 192
    vehicle, 173, 176, 177
  efflux pumps, 8
  nanocarriers, 9
  permeation, 19
  release, 2, 5, 7–9, 12, 13, 17–19, 21, 23, 27, 29, 50, 61–63, 74–77, 79–85, 89–96, 98, 101, 103–107, 153, 156, 163, 180, 186, 187, 191–193, 198, 220, 223, 224, 227
dynamic, 2, 163
  light scattering, 52, 53, 106
  mechanical analysis (DMA), 207
  physicochemical alterations, 26

**E**

elastin-mimetic hybrid
  polymers (EMHPs), 203
elastomer, 9, 24
electrochemical response, 162
  targeting homeostasis, 163
electromagnetic radiations (EMR), 92
electron
  beam, 212
  microscopy, 206
  transport chain, 137
  withdrawing groups, 16
electrostatic interaction, 8, 190
emulsification, 99, 106
endocytosis, 14, 103, 104, 153, 187, 190
endogenous improved green fluorescence protein, 26, 32
endopeptidases, 50, 140

endosomal
  environment, 135
  escape, 101, 103, 127, 129, 130, 139, 178
  membrane, 178
  pathway, 178
endosomes, 7, 90, 101, 102, 122, 189
endothelial cells, 93, 158, 188, 208
enhanced
  green fluorescent protein (EGFP), 26, 137
  permeability and retention (EPR), 152, 153, 191
environmental signals, 1, 220
enzymatic
  digestion, 12, 50, 81
  oxidation, 30, 48
  reaction, 15, 59
  triggers, 12
enzyme
  dysregulations, 12, 50
  nanoparticles activation, 190
  responsive polymers, 5, 140
epichlorohydrin (ECH-HPP), 222
epithelial growth factor receptor (EGFR), 104
Epstein Barr virus (EBV), 122, 144
equilibrium, 19, 42–46, 56, 58, 163
ester bonds, 12, 50, 131, 226
esterification, 200
ethylene glycol dimethacrylate (EGDMA), 77, 108
European
  Randomized Study of Screening for Prostate Cancer (ERSPC), 7
  Science Foundation (ESF), 152
extracellular
  environment, 7, 20, 135
  fluids, 6, 10, 191
  matrix (ECM), 101, 163, 166, 180, 198, 202, 218, 229
  microenvironment, 13

**F**

finite element modeling (FEM), 165, 166
first-pass metabolism, 85, 86
fluorescein isothiocyanate dextran (FITC-dextran), 228

fluorescence
  indication, 17
  microscopy, 99
  response, 227, 228
  spectroscopy, 51, 106
fluorophore, 157, 158, 227
Food and Drug Administration (FDA), 7,
  8, 165, 191, 201
Fourier transform infrared (FTIR), 51, 63
free radical polymerization, 42, 45, 209,
  212, 213
functional groups, 5, 10, 31, 41, 44, 46, 51,
  184, 185, 197, 210–213, 221

**G**

gel permeation chromatography (GPC),
  50–52, 216
gelatin corona, 103
gelation, 15, 28, 29, 55, 56, 86, 88, 89, 95,
  96, 165, 200, 223
  process kinetics, 56
  temperatures, 15
gene
  delivery, 8, 10, 13, 123, 126, 133, 136,
    139, 142, 143, 154, 156
    bioresponsive polymers, 124
    pH-responsive polymers, 124
    reductive environment responsive
      polymers, 134
    systems, 123, 156
    thermoresponsive polymers, 131
    vectors, 135
  expression, 126, 130, 135, 142, 219
  silencing, 26, 27, 90
  therapy, 27, 121–123, 143, 156, 165
  transfection, 101, 133, 134, 137, 139
  transfer, 122, 123, 130, 135
glucagon-like peptide-1 (GLP-1), 84, 108
gluconic acid, 16, 17, 31, 48, 59, 60, 78, 162
glucose
  binding proteins, 17
  molecule, 31, 60
  oxidase (GOx), 15–18, 30, 31, 48, 59,
    60, 78, 161, 162
  responsive
    insulin delivery system (GRIDS),
      15, 16

material, 16
swelling, 16
sensitive
  enzyme, 17
  nanoparticle systems, 17
  platforms, 18
  polymers, 5
  system, 31
signal amplifiers, 78
glucuronic acid, 96, 200
glutaraldehyde, 30, 200, 211
glutathione (GSH), 6, 10, 12, 49, 62, 124,
  134–137, 144, 191, 192, 224, 225
glyceraldehyde 3-phosphate dehydrogenase (GAPDH), 90
glycoproteins, 154
glycosaminoglycan (GAG), 200
gyrase subunit B (GyrB), 84

**H**

heparin, 13, 14, 79, 80, 99, 183
hepatic metabolism, 86, 87, 89
hepatitis B virus (HBV), 122, 144
hepatocytes, 143, 181, 182
herpes simplex virus (HSV), 122
heterogeneity, 102, 179, 202
high
  density lipoproteins (HDL), 179
  molecular weight (HMW), 47, 107, 142
Hofmeister series, 19
homeostasis, 161, 164, 225
homopolymers, 24, 45, 203
human
  immunodeficiency virus (HIV), 106,
    122, 134, 174
  neutrophil elastase (HNE), 75, 108
  renal carcinoma cell line, 141
  serum albumin, 84
  umbilical vein endothelial cells
    (HUVECs), 130
hyaluronic acid (HA), 13, 14, 16, 21, 61,
  74, 75, 78–80, 138, 183, 189, 190, 193,
  199, 200, 202, 221, 224
hyaluronidase (HAase), 6, 7, 13, 14, 21,
  22, 75
  -1 (Hyal-1), 14

hybrid hydrogels, 199, 202
hydrogel, 2, 9, 13, 16, 24, 25, 56, 59–63,
74, 75, 81, 86, 87, 89, 94–96, 98, 100,
133, 158–165, 197–214, 216–221, 223,
224, 226–229
    classification based, 198
        configuration, 204
        crosslinked chain charge, 204
        crosslinking, 204
        origin, 199
        physical appearance, 204
        polymeric composition, 203
    introduction, 197
    matrix, 159, 165, 207
    physical and chemical properties, 205
        biocompatible properties, 207
        mechanical properties, 207
        porosity, 206
        swelling properties, 205
    systems, 14, 58, 76, 77, 202, 227
hydrogen
    bonding, 27, 30, 51, 209, 210
    peroxide, 10, 49, 78, 79, 137, 225
hydrolysis, 103, 130, 139, 141, 200
hydrophilic
    carriers, 103
    components, 23
    groups, 84, 205
    lipophilic-balance (HLB), 213
    moieties, 77, 80, 133
    molecules, 87, 98
    segments, 29
hydrophilicity, 26, 49, 125, 182–185
hydrophobic
    agents, 23
    backbone, 62
    block, 55
    constituents, 16
    core, 179, 189
    drugs, 11, 15, 100
    groups, 24, 28, 30, 205
    interactions, 30, 131, 204, 207, 211
    moiety, 30
    molecules, 84
    monomers, 24
    polymer backbone, 41
    segments, 29, 133, 211

hydrophobicity, 179, 182, 184, 185
hydroxyethyl
    cellulose/hyaluronic acid (HECHA), 74
    methacrylate (HEMA), 24, 76, 77, 93,
    162, 201, 202
hydroxyl
    groups, 10, 31, 226
    propyl methylcellulose (HPMC), 80, 87,
    96, 97, 134
    radical (OH), 9, 10, 49, 80, 131, 137,
    157, 166, 211, 212, 225
hypochlorous acid (HOCl), 137, 157, 225
hypoxia
    inducible factor 1a (HIF-1a), 189
    nanoparticles activation, 188
    responsive nanoparticles (HR-NPs), 23,
    189, 190
hypoxic
    cells, 23
    conditions, 23, 78, 189
    regions, 189
    tumor tissues, 23

**I**

immune cells, 155, 173, 176, 177
immunogenicity, 90, 123, 126, 130, 136,
    143, 152, 154, 166, 174, 200
immunoglobulins, 182
immunotherapeutics, 103
immunotherapy, 75
*in situ*
    gel, 86, 89
    gelation, 19, 88
    gelling ocular system, 97
    nasal systems, 85
*in vitro*
    cytotoxicity, 23
    drug release, 29
    transfection, 103
*in vivo*
    conditions, 81
    imaging, 22
    response, 75
    sol-gel transition, 29
    studies, 78, 86, 103
    tests, 100

toxicity, 103, 207
xenograft tumor model, 21
indomethacin, 76, 94, 95, 192
inflammatory
  bowel diseases, 12
  mediators, 76
infrared (IR), 22, 51, 63, 92, 96, 97, 215, 216
inhomogeneities, 50, 53
intelligent polymers, 27
intermolecular
  bonding, 73
  hydrogen bonds, 9
  organization, 2
interpenetrating polymeric hydrogel, 204, 229
interstitial fluid, 19, 77, 180
intracellular
  acid, 12, 50
  compartments, 19
  hydrolysis, 21
  matrix, 62
  microenvironments, 101
  site, 26
  trafficking, 180, 192
  vesicles, 122
inverse-suspension polymerization, 213
ionic
  concentration, 18, 19
  strength, 2, 18–20, 57, 62, 74, 208, 221
ionizable groups, 7, 29, 58, 62, 83
ionization, 29, 55, 57, 59, 61, 62, 127
iontophoresis, 76, 77
ischemia, 6, 22
isoelectric point, 55, 81, 210

**K**

keratoconjunctivitis, 97
Ketorolac tromethamine nasal sprays, 86
Kupffer cells, 182

**L**

*Lactobacillus*
  *acidophilus*, 154
  *crispatus*, 106
*Laminaria hyperborea*, 200

lectin, 15, 30, 31, 48
levofloxacin, 96, 97, 224
ligands, 19, 43, 123, 181, 187
lipophilic
  balances, 27
  drugs, 29
  moieties, 28
lipoplexes, 124, 178, 181
liposomes, 100, 103–105, 123, 177, 178
liquid chromatography, 52
living radical polymerization (LRP), 42, 43, 45
low
  density lipoproteins (LDL), 179, 193
  frequency ultrasound (LFUS), 104, 105
lower critical solution temperature
  (LCST), 19, 23, 24, 28, 51, 85, 99, 131, 133, 134, 201, 223, 224
lymphocytes, 61, 155
lysosome, 6, 7, 90, 122, 124, 178

**M**

macromolecular
  backbones, 212
  drugs, 165
  structures, 161
macromolecules, 45, 52, 158, 199, 206, 216
macrophages, 61, 154, 155, 182, 188, 225
macro-radicals, 212
magnetic
  field, 2, 74, 85, 92, 208
  mesoporous silica (MMS), 17, 18
  resonance imaging (MRI), 9, 191
maleic acid amide (MAA), 11, 81–83, 89, 129, 130
mammary-derived growth inhibitor
  (MDGI), 187
mannan
  and chitosan co-modified PLGA
    (MN-CS-PLGA), 90
  modified PLGA (MN-PLGA), 90
matrices, 153, 159, 162, 199, 200, 202, 220
matrix metalloproteinase (MMP), 6, 7, 12, 14, 15, 50, 103, 104, 140, 142, 190, 191, 193, 202, 229
maturation, 177, 211

mesenchymal transition, 22, 188
mesoporous silica, 9, 11, 14, 21, 84, 103,
    174
    nanoparticles (MSPs), 11, 14, 84, 103,
    104, 174
metastasis, 12, 13, 22, 188
methacrylate, 7–9, 30, 44, 60, 76, 80, 93,
    128, 129, 132–134, 202, 221, 222
methacrylic acid, 4, 7, 80–83, 106, 203
methacryloyloxyethyl phosphorylcholine
    (MPC), 192, 203
methicillin-resistant *Staphylococcus*
    *aureus* (MRSA), 175
methotrexate (MTX), 185
methoxy polyethylene glycol-b-
    poly(diethyl sulfide), 139
micellar structures, 50, 54
micelles, 9, 19, 50–55, 84, 102, 107, 127,
    128, 132, 133, 135, 139–141, 156, 178,
    187, 192, 193, 201, 211, 221, 224
micellization, 55, 102
Michael addition reaction, 136
Michaelis-Arbuzov reaction, 207
micro electro mechanical system (MEMS),
    92
microcapsules, 21, 61, 85
microenvironment, 6, 76, 77, 101, 142
microgels, 8, 228
microneedles, 77, 78, 79
microorganisms, 124, 199
microparticles, 8, 49, 82, 89, 90, 103, 186,
    226
microRNA (miRNA), 26
microspheres, 8, 11, 24, 77, 80, 81, 87, 90,
    204, 213
molecular
    drugs, 174
    gates, 11
    imaging, 156, 157
    oxidants, 137
    pathways, 156
    size, 51, 52
    species, 137, 225
    structure, 29, 48
    target, 157
    weight, 30, 42, 43, 45, 47, 50, 52, 98,
    128, 153, 201, 211, 216, 224

Moloney murine leukemia virus
    (MoMLV), 122
monomers, 15, 17, 43–48, 57, 58, 61, 80,
    85, 93, 99, 127, 130, 135, 163, 193, 198,
    203, 208, 209, 211–213
mononuclear phagocytic system (MPS),
    180, 182, 183
morpholinoethane sulfonic acid (MES),
    203
mutation, 121, 175

**N**

N-(2-hydroxypropyl) methacrylamide
    (HPMA), 136, 140, 152, 153, 202
N-acryloyl pyrrolidine (APy), 24
nanocarriers, 9, 13, 17, 21, 23, 75, 83, 90,
    103, 104, 136, 158, 188–190
nanocomplexes, 83, 84, 143
nanomaterials, 180, 181
nanomedicines, 151, 152, 173, 175,
    179–183, 187, 192
nanometer, 53, 152
nanoparticles (NP), 9, 11, 12, 17, 21, 23,
    24, 81, 82, 84, 90, 101, 104–106, 129,
    132, 133, 136–139, 154, 155, 173, 174,
    177–185, 187–193
nanoscale, 18, 41, 125
nanosystem, 104, 142
nanotherapeutic, 181
    strategy, 104
    treatment, 104
nasal
    administration, 86, 88
    cavity, 86–88, 90
    drug delivery systems, 85
natural hydrogels, 199–201
near-infrared (NIR), 22, 92, 93
neuroblastoma cells, 188
next-generation
    delivery systems, 27
    precision medications, 1
N-hydroxysuccinimide (NHS), 203
N-isopropylacrylamide (NIPAMM), 46,
    93, 94, 134, 201, 202, 223, 224
nitric oxide, 137, 225
nitroimidazole (NI), 16, 23, 78, 189, 190

nitroxide-mediated radical polymerization
  (NMRP), 41
N-methyl-2-pyrrolidone (NMP), 98
nonviral gene
  delivery, 27
  vectors, 103, 156
norfloxacin, 93, 96
normoxic
  cells, 23
  conditions, 23
novel
  biodegradable polymers, 1
  macromolecular platform, 107
  triblock polymeric system, 29
nuclear
  compartment, 101
  magnetic resonance (NMR), 50, 51, 54,
    205
  scattering, 54
nucleic acids, 2, 8, 26, 27, 48, 79, 101,
    103, 122–125, 129, 133, 135, 140–142,
    158, 161, 184, 208
  ROS responsive polymers, 137
nucleotides, 8, 21, 141

**O**

oligodeoxynucleotides (ODNs), 122
oligonucleotide, 27, 123
opsonization, 180, 182, 183, 189
organic
  compounds, 226
  materials, 8
  solvent, 98, 226
orthoester groups, 125
osmotic
  effects, 206
  pressure, 29, 93, 126
oxidants, 10, 137, 225
oxidative stress, 11, 137, 225–227
oxygen, 6, 7, 9, 10, 16, 42, 43, 46, 49, 61,
    76, 79, 105, 137, 139, 157, 186, 225

**P**

paclitaxel (PTX), 9, 15, 98, 102, 187, 190,
    191
particle replication in non-wetting
  templates (PRINT), 8

pathophysiology, 152, 179
payload molecules, 8, 21
pectin, 13, 20, 49, 60, 81, 89, 140
peptides, 19, 48, 75, 77, 79, 81, 92, 94, 95,
    98, 123, 132, 140, 154, 158, 161, 177, 184,
    187, 188, 192, 202, 203, 208, 219, 220
peroxyl radicals, 137, 225
peroxynitrite (ONOO-), 137, 157, 166, 225
pH
  gradient, 7, 124
  responsive
    degradable structure, 128
    hydrogels, 61, 220
    materials, 5, 8
    protonatable structures, 125
  sensitive
    micelles, 9
    polymers, 17, 29, 80
    pyridine surface, 77
  value, 16
phagocytic cells, 182
phagocytosis, 155, 180
pharmacokinetics, 84, 85, 90, 107
phenylboronic acid (PBA), 10, 16–18, 30,
    31, 48, 49, 60, 141
phosphate-buffered saline (PBS), 28, 136,
    137
phosphatidylcholine, 100, 187
phospholipase A2 (PLA2), 6, 151
phosphoric acids, 221
photodynamic
  effect, 105
  therapy (PDT), 93, 105, 190, 193
photon correlation spectroscopy, 52
photosensitizer, 105, 142
physicochemical
  characterization, 166
  factors, 83
  properties, 124, 135, 180, 182, 184, 185
physiological temperature, 23, 100, 223
plasma, 2, 3, 6, 13, 87, 88, 100, 101, 107,
    122, 124, 152, 182, 183, 187, 225
plasmid, 122, 129, 130, 133, 137
  DNA (pDNA), 122, 126, 127, 129, 130,
    132, 134, 138, 140, 142
pluronics, 28, 134
polo-like kinase 1 (PLK-1), 141

poloxamer, 4, 86, 87, 88
poly(2-acrylamido-2-methylpropane
   sulfonic acid) (PAMPS), 221
poly(4-styrenesulfonic acid) (PSSA), 221,
   222
poly(acrylic acid) (PAA), 29, 31, 75, 96,
   135, 136, 139, 201, 210
poly(amidoamine) (PAMAM), 20, 26, 83,
   84, 103, 136
poly(dimethyl amino ethyl methacrylate)
   (PDMAEMA), 30, 127–129, 132, 134
poly(lactic-co-glycolic acid) (PLGA), 8,
   11, 24, 28, 77, 90, 94, 98, 106, 138, 154,
   155, 173, 174
   nanoparticles, 11, 154, 174, 177
poly(L-lysine) (PLL), 125, 131, 132, 135,
   137, 138, 140, 141, 143, 156
   graft-chitosan (PLL-g-Chi), 137
poly(lysine-cholic acid) (P(Lys-Ca)), 9
poly(methacrylic acid) (PMAAc), 29
poly(MPC-co-methacrylic acid) (PMA),
   83, 203
poly(*N*, *N*-diethylamino-ethyl methacrylate) (PDEAEMA), 30
poly(*N*, *N*-dimethylamino-ethyl methacrylate) (PDMAEMA), 30
poly(N-isopropyl acrylamide)
   (PNIPAAM), 24, 28, 84, 85, 93, 94, 131,
   133, 134, 224, 227
poly(N-vinylpyrrolidone) (PVP), 22, 87
poly(propylene sulfide) (PPS), 10, 48, 49,
   227
poly(styrene-alt-maleic anhydride)
   (pSMA), 132
poly(β-amino ester) (PBAE), 90, 130, 132,
   187
poly(ε-caprolactone) dimethacrylate
   (PCLDMA), 224
poly[(2-diethylamino) ethyl methacrylate]
   (PDEA), 221
poly[(2-diisopropylamino) ethyl methacry-
   late] (PDPA), 132, 221, 222
poly[(2-N-morpholino) ethyl methacrylate]
   (PMEMA), 221, 222
polyamidoamine (PAMAM), 131, 136,
   189–191
polyaniline, 94, 95, 162

polyanion, 59, 60, 129, 207, 210
polyaspartamide (PAsp), 125, 127, 128,
   132, 138, 140
polycations, 8, 27, 59, 91, 123, 128–130,
   136, 142, 207, 210
polydispersities, 43, 45, 50
polydispersity index (PDI), 43
polyelectrolytes, 29, 62, 78, 220
polyesters, 130, 212
polyethylene
   glycol (PEG), 9–11, 28, 29, 31, 44, 48,
      49, 56, 60, 75, 78, 82, 83, 86, 88, 89,
      94, 100–103, 106, 126, 127, 129, 132,
      137, 139–141, 174, 178, 179, 183, 187,
      189–193, 201–203, 210, 211, 223, 228
   oxide (PEO), 55, 84, 100, 210
polyethyleneimine (PEI), 103, 125–127,
   131–133, 135, 137, 138, 143, 189, 190
   nitroimidazole micelles (PEI-NI), 189, 190
polyion complex micelles, 8, 19
poly-l-glutamate (PLG), 126
polylysine (PL), 49
polymer
   enzyme liposome therapy (PELT), 151
   macromolecules, 50
   matrices, 2
   micellar solutions, 50
   network, 57, 62, 227
   shell gate, 12
   solutions, 15, 28, 51
   therapeutics, 152
polymeric
   backbone, 102, 197
   bases, 30
   carriers, 26, 123, 140
   gene complexes, 142
   materials, 49, 73, 140, 225
   matrices, 57, 197
   membrane, 76
   micelle packing, 28
   monophasic system, 28
   nanocarriers, 13, 107
   nanogel, 21
   network, 2, 30, 203
   ratio, 94
   systems, 49
   transporters, 26

polymerization, 11, 41–47, 50, 83, 127, 130, 193, 198, 200, 203, 204, 208, 212, 213, 227
polymerized crystalline colloidal array (PCCA), 228
polymerprotein, 153, 156
polymersomes, 16, 23, 49, 133, 178
polymethacrylic acid, 60, 89
polymorphonuclear (PMN), 61
polypeptides, 47, 201
polyplexes, 26, 27, 79, 101, 123–131, 135, 136, 138–140, 142, 156, 178
polysaccharides, 2, 13, 14, 20, 28, 60, 73, 82, 137, 139, 183, 199, 200, 212, 222
polystyrene, 4, 19, 48, 229
polythiophene, 76, 162
polyvinyl alcohol (PVA), 76, 201, 211, 229
porous silicon (PSi)
    carriers, 184, 185
    drug conjugates, 185
    films, 186
    framework, 185
    materials, 184–186
    matrix, 184–186
    nanoparticles, 184, 185, 187, 188
    nanovectors, 188
    particles, 183–187
    surface, 185
prodrug, 191, 193
programmed death-1 (PD-1), 75
*Propionibacterium acnes*, 74
prostate,
    lung, colorectal, and ovarian cancer screening trial (PLCO), 7
    specific antigen (PSA), 7, 12, 50
protein kinase Cα (PKCα), 13
proton, 29, 126, 220, 221
    pump inhibitors, 8, 81
    sponge, 125–127
protonation, 5, 26, 31, 60, 62, 124–127, 132, 162, 220, 221

**Q**

quadruple stimuli-responsive system, 142
quasi-elastic light scattering (QELS), 52, 53
quaternary ammonium, 19, 29, 139

**R**

radical polymerization, 43, 45, 127, 224
radiosensitizer, 105
reactive oxygen species (ROS), 6, 10, 11, 49, 82, 105, 137, 139, 140, 142, 157, 192, 193, 225–227
red blood cells (RBCs), 154, 155
redox
    condition, 192
    couple, 10
    enzyme, 163
    nanoparticles activation, 191
    potential, 10, 134
    properties, 76
    sensitive materials, 10
    stimuli-responsive gene delivery systems, 138
reducible hyperbranched (rHB), 26, 138
regional hyperthermia, 105, 106
renal filtration system, 182
reticuloendothelial system (RES), 9, 154, 182
rheological properties, 56, 86
ring-opening metathesis polymerization (ROMP), 42
RNA interference (RNAi), 26
ruthenium, 43, 47

**S**

scanning electron microscope (SEM), 99, 215
scattering techniques
    laser light scattering, 52
    small-angle x-ray and neutron scattering, 53
silica nanoparticles, 180, 189
silicon (Si), 16, 183, 184
single-walled carbon nanotubes (SWNTs), 17, 163, 166
small
    angle
        neutron scattering (SANS), 53, 54
        scattering (SAS), 50, 53, 54
    interfering RNA (siRNA), 22, 26, 82, 87, 90, 101, 122, 130, 132, 138, 139, 141, 142, 178, 181, 189

sol-gel transition, 28, 56, 100, 223
spectroscopy, 51, 106
stimuli-responsive systems, 103, 108, 142, 186
streptozotocin diabetic rat model, 107
sulfonic
  acid, 19, 62, 221
  group, 209
superoxide
  dismutase (SOD), 137
  radicals, 137, 225
supramolecular
  chemistry, 166
  gels, 9
  network, 9
  systems, 19
surfactants, 28, 51, 98
suspension polymerization, 213

**T**

T cell, 103, 155
tellurium oxidation, 226
therapeutic, 8, 16, 22, 26, 27, 50, 133, 139, 153–155, 163, 165, 166, 180, 181, 223
  agent, 1, 93, 121, 124, 131, 152, 186, 207, 220
  concentration, 27
  devices, 20
  dose, 2, 104
  effect, 180, 181
  efficacy, 96, 104, 105, 180, 181
  gene delivery, 132
  molecules, 21
  nucleic acid, 26, 121, 125
  protein, 8
  range, 2, 3
  targets, 13
thermodynamic, 159
  equilibrium, 57, 58
  relation, 28
thermoresponsive
  biopolymers, 88
  drug carrier, 24
  gels, 86
  hydrogels, 223
  micelles, 133

polymers, 84, 133, 223
  segments, 133
  systems, 86
  transition course, 24
thermosensitive
  carriers, 133
  hydrogels, 211
  polymeric
    solution, 28
    systems, 29
  polymers, 28
  systems, 29
thioketal nanoparticles (TKNs), 82, 139
thiol groups, 10, 49, 135
thiolation, 136
thioredotoxin (Trx), 134
three-dimensional
  network, 88
  spheroid cultures, 23
  structure, 73
threshold, 6, 154, 158
thrombin, 13, 14, 79
tissue
  engineering, 1, 107, 153, 159, 163, 166, 199, 202, 207, 208, 217–219, 224
  regeneration, 9, 198, 217, 229
transmission electron microscopy, 106
transnasal drug delivery, 91
triblock copolymer, 135
triglycerides, 179
trimethyl-locked benzoquinone (TMBQ), 10
tumor
  microenvironment (TME), 12, 13, 21, 103, 131, 180, 189–192
  necrosis factor
    alpha (TNF-α), 139
    related apoptosis-inducing ligand (TRAIL), 13, 14

**U**

ultrasound, 2, 4, 74, 85, 91, 104
ultraviolet (UV), 51, 92, 93, 200, 207, 213
upper critical solution temperature (UCST), 28, 131, 223, 224

**V**

vancomycin, 175
vascular
  diseases, 22
  endothelial growth factor (VEGF), 93,
    144
vesicle bilayers, 55
vinyl
  acetate, 4, 202
  monomers, 46
virosomes, 154
viscoelastic properties, 56
viscosity, 56, 85, 86, 88, 97, 99, 185, 213,
    215
vitreoretinopathy, 92

**W**

Washburn's equation, 206
water-in-oil (W/O), 213

**X**

xanthan, 89, 210, 211
xenograft, 21, 106
x-ray
  diffraction, 216
  scattering, 53
xyloglucan, 4, 88
  polysaccharide, 100

**Z**

zeta potential, 133, 179

Printed and bound by CPI Group (UK) Ltd, Croydon, CR0 4YY

23/10/2024

01777702-0003